Ilya Charaev

Improving the Spectral Bandwidth of Superconducting Nanowire Single-Photon Detectors (SNSPDs)

Karlsruher Schriftenreihe zur Supraleitung Band 020

HERAUSGEBER

Prof. Dr.-Ing. M. Noe
Prof. Dr. rer. nat. M. Siegel

Eine Übersicht aller bisher in dieser Schriftenreihe erschienenen Bände finden Sie am Ende des Buches.

Improving the Spectral Bandwidth of Superconducting Nanowire Single-Photon Detectors (SNSPDs)

by
Ilya Charaev

KIT Scientific Publishing

Karlsruher Institut für Technologie
Institut für Mikro- und Nanoelektronische Systeme

Improving the Spectral Bandwidth of Superconducting
Nanowire Single-Photon Detectors (SNSPDs)

Zur Erlangung des akademischen Grades eines Doktor-Ingenieurs
von der KIT-Fakultät für Elektrotechnik und Informationstechnik des
Karlsruher Instituts für Technologie (KIT) genehmigte Dissertation

von Ilya Charaev, geboren in Russland

Tag der mündlichen Prüfung: 14. Dezember 2017
Erster Gutachter: Prof. Dr. rer. nat. Michael Siegel
Zweiter Gutachter: Prof. Dr. rer. nat. Heinz-Wilhelm Hübers

Impressum

KIT Scientific Publishing

Karlsruher Institut für Technologie (KIT)
KIT Scientific Publishing
Straße am Forum 2
D-76131 Karlsruhe

KIT Scientific Publishing is a registered trademark
of Karlsruhe Institute of Technology.
Reprint using the book cover is not allowed.

www.ksp.kit.edu

Print on Demand 2018 – Gedruckt auf FSC-zertifiziertem Papier

ISSN 1869-1765
ISBN 978-3-7315-0745-1
DOI 10.5445/KSP/1000077944

Kurzfassung

Diese Arbeit stellt eine umfassende Untersuchung des Einflusses von geometrieabhängigen Faktoren auf die Leistungsmerkmale von supraleitenden Einzelphotonen-detektoren (Superconducting Nanowire Single-Photon Detectors, SNSPD) dar. SNSPDs basieren auf einem supraleitenden Film, welcher zu einer mäanderförmigen Struktur im Nanometer Bereich gefertigt wurde. Wird diese Struktur von einem Strom (Biasstrom) knapp unterhalb des kritischen Stroms durchflossen, so reicht die Absorption eines einzelnen Photons um die Supraleitung lokal zu zerstören, wodurch ein messbarer Spannungspuls entsteht. Die Eigenschaft Licht im Einzelphotonenbereich zu detektieren ist für viele Anwendungen interessant, insbesondere wenn es sich um äußerst schwache Lichtsignale handelt. SNSPDs entwickelten sich daher schnell als vielversprechende Photonen- zähltechnologie für optische und infrarote Wellenlängen. Sie besitzen eine hohe Detektion- seffizienz, eine geringe Dunkelzählrate und eine ausgezeichnete zeitliche Auflösung. Die Detektionseffizienz nimmt jedoch exponentiell nach Überschreiten eines bestimmten Schwellenwerts ab, der hauptsächlich durch den begrenzten experimentellen kritischen Strom bestimmt ist. Speziell im infraroten Bereich ist die Effizienz daher stark beeinträchtigt. Das Ziel dieser Arbeit ist somit die spektrale Bandbreite der SNSPDs in den infraroten Bereich zu erweitern und die theoretisch vorhergesagte Abhängigkeit des Schwellwerts experimentell zu bestätigen.

Um die supraleitenden Eigenschaften der Detektoren, im speziellen des kritischen Stroms, zu verbessern wurde der Strukturierungsprozess optimiert. Dies erfolgte durch die Entwicklung eines Negativ-Elektronenstrahl-Lithographie-Prozesses unter Verwendung von positivem Photolack, welcher viele Vorteile für die Herstellung von SNSPDs bietet. Zusätzlich wurde ein zweischichtiges AlN / NbN-System verwendet, um den Proximity-Effekt in den Nanodrähten zu unterdrücken. Um den sog. current crowding Effekt zu verhindern und die gleichmäßige Stromverteilung zu verbessern, wurde ein knickfreies SNSPD Design in Form einer Spirale verwendet. Die dafür notwendige Mehrschicht- technologie für die Herstellung der Spiral-SNSPDs wurde ebenso im Rahmen dieser Arbeit entwickelt.

Um den 3D-Proximity-Effekt in Nanodrähten zu beschreiben, wurde ein laterales NSN-Proximitätsmodell entwickelt. Die experimentellen Ergebnisse, welche durch Messung von NbN-Streifen einer Breite zwischen 50 nm und 20 μm gewonnen wurden, werden dabei allgemein von diesem NSN Modell beschrieben. Weiterhin wird ein Ansatz vorgeschlagen um die Breite des supraleitenden Bands abhängig vom magnetischen Feld zu bestimmen.

Um den Einfluss des experimentellen kritischen Stroms auf den Schwellenwert zu untersuchen, wurden die für supraleitende Einzelphotonendetektoren relevante Leistungs- parameter experimentell an drei verschiedenen SNSPD-Geometrien (Mäander, Doppel- spirale und Einzelspirale) untersucht. Die Spektren der Detektoreffizienz wurde dabei im Wellenlängenbereich von 400 nm bis 1600 nm gemessen und jeweils ihr Schwellwert bestimmt, hinter dem die Effizienz der Detektoren abfällt. Die größte und die kleinste Schwelle wurde dabei für das Einzelspiral-Layout (943 nm) bzw. für den Mäander (605 nm) gefunden. Für alle drei Layouts fällt die Beziehung zwischen dem Schwellwert und dem relativen Biasstrom auf eine universelle Kurve, die im Rahmen des modifizierten Hot-Spot-Modells vorhergesagt wurde. Die Tatsache, dass alle experimentellen Punkte auf die gleiche theoretische Kurve fallen, zeigt die Schlüsselrolle des experimentellen kritischen Stroms in Bezug auf das abknicken der Detektionseffizienz bei höheren Wellenlängen. Die Tatsache bestätigt ebenso die Annahme, dass die Schwelle durch die positionsabhängige Wahrscheinlichkeit von Photonen verursacht wird, in einer geraden Linie detektiert zu werden, und die Schwelle nahezu unbeeinflusst ist von den Biegungen im Layout. Bei der Messung des Einzelspiral-Designs zeigte sich die erwartete höchste Detektionseffizienz für Wellenlängen unterhalb des Schwellwerts und der Zeit-jitter pro Längeneinheit des Nanodrahts zeigte den kleinsten Wert. Im Gegensatz dazu sind sowohl die Dunkelzählrate als auch der Timing-Jitter durch die Qualität und Anzahl der Biegungen im Layout beschränkt. Es zeigte sich, dass der Wert des Timing-Jitters direkt von der Länge des Nanodrahts, seiner kinetischen Induktivität und den Biegungen abhängt.

Durch die Verwendung spezieller Strukturen in Form einer quadratischen Spirale, bei der alle Biegungen die gleiche Symmetrie in Bezug auf die Richtungen des Biasstrom und des externen Magnetfeldes aufweisen, wurde gezeigt, dass die Bereiche um die Biegungen bei geringen Photonenenergien wesentlich zur Dunkelzählrate, sowie zur Photonenzählrate beitragen.

Abstract

This work presents a comprehensive investigation of the influence of geometry-dependent factors on performance metrics of superconducting single-photon detectors (Superconducting Nanowire Single-Photon Detector, SNSPD). SNSPDs are based on a superconducting film nanofabricated into a meandering wire geometry. They are biased at a current just below the critical level. The absorption of a photon in the superconductor causes a local collapse of superconductivity, whereby a measurable voltage signal is detected. The ability to detect light on a single-photon level is attractive for many applications especially where the detection of low light intensity is required. SNSPDs rapidly emerged as a highly promising photon-counting technology for optical and infrared wavelengths. They have high efficiency, low dark counts and excellent timing resolution. However, the detection efficiency exponentially decreases above a certain threshold which is limited mainly by the restricted experimental critical current. This greatly impairs the performance in the infrared range. The aim of this thesis is to extend the spectral bandwidth of SNSPD towards infrared range. At the same time, the theoretical predicted dependence of threshold has to be confirmed experimentally.

In order to enhance the superconductivity in nanowires, in particular the critical current, the quality of patterning process has been improved. This is done by developing a negative electron-beam lithography process on positive resist. A whole range of advantages which is offered by negative lithography with positive PMMA resist ensures its high potential for improving performance metrics of superconducting nanowire singe-photon detectors. Additionally, a bi-layer AlN/NbN system was used to suppress the proximity effect in nanowires. To avoid current crowding and improve the uniformity of the current distribution in a detector, a bend-free layout was used as a geometry of SNSPD. A multi-layer technology for single-spiral bend-free SNSPD has been developed.

To describe the 3D proximity effect in nanowires, a lateral NSN proximity model has been developed. The experimental results which were obtained on thick NbN strips with widths in the interval from 50 nm to 20 μm, are universally described by the NSN model. Furthermore, an approach for the determination of the width of the superconducting band from the value of magnetic field has been proposed.

To investigate the role of the experimental critical current in position of the threshold, the performance parameters, relevant for nanowire single-photon detectors, have been experimentally evaluated for three different layouts of bended nanowires (meander, double spiral, single spiral). The spectra of the single-photon response have been measured in the wavelength range from 400 nm to 1600 nm and defined the threshold beyond which the response rolls off. The largest and the smallest threshold were found for the single-spiral layout (943 nm) and for the meander (605 nm), respectively. For all three layouts the relationship between threshold and the relative bias current falls onto a universal curve which has been predicted in the framework of the modified hot spot model. The fact that all experimental points fall on the same theoretical curve evidences the key role of the experimental critical current in the position of the spectral cut-off. This fact also confirms the supposition that the threshold is caused by the position dependent probability of photons to be detected in a straight line and that it is almost unaffected by the bends in the layout. For the single-spiral layout, the efficiency of photon detection at wavelengths smaller than threshold reaches the expected absorbance of the spiral structure and the timing jitter per unit length of the nanowire has the smallest value. Contrary, the rate of dark counts as well as the timing jitter are mostly restricted by the quality and number of bends in the layout. The value of timing jitter is directly related to the length of the nanowire, its kinetic inductance and the bends.

Using special structures in the form of a square spiral, where all bends have the same symmetry with respect to the directions of the bias current and external magnetic field, it was shown that the areas around the bends largely contribute to the rate of dark counts and to the rate of photon counts at small photon energies. Invoking handedness of the observed asymmetries and mapping the computed local absorption probability for photons and the local detection threshold current, areas in the bends where, at low photon energies, photon counts occur is identified. The microscopic scenario of the photon count event explains the faint effect of the inverted asymmetry in the count rate for low-energy photons at small fields and currents. At large magnetic fields and currents, the asymmetry in the rate of photon counts disappears as it is predicted by the theoretical model. Asymmetry in count rates is absent in circular spirals without bends. Furthermore, the absorption of spiral detectors is independent on polarization. This allows to integrate the detector easier in optical system.

Contents

Preface

The work presented in this dissertation has been carried out at the *Institute of Micro- and Nanoelectronic Systems* (IMS) at the *Karlsruhe Institute of Technology* (KIT). This research was realized within the framework of the program “Optics & Photonics” offered by *Karlsruhe School of Optics and Photonics* (KSOP). I want to thank the heads of both institutes, Prof. Siegel and Prof. Lemmer for offering me the opportunity to perform the research.

These four years have been a challenging trip, with both ups and downs. I would have never reached the point of finishing my dissertation without the help and support of others. For this, I would like to kindly thank them.

First of all, I would like to thank my supervisor Dr. Konstantin Ilin for his excellent guidance and support during this process. I deeply appreciate all the advice, insightful comments, and kind words he has given me. Konstantin’s constant enthusiasm for research fuelled me during my time at IMS, and I suspect that most of the good ideas I had were secretly planted there by Konstantin, and he convinced me they were my own.

I especially thank Prof. Semenov from DLR for helpful discussions and successful collaboration.

My colleagues have contributed immensely to my personal and professional time at IMS. Our group has been a source of friendships as well as good advice and collaboration. I am grateful to my colleagues, Steffen Dörner, Artem Kuzmin, Michael Merker, Ekkehart Schmidt for the scientific exchange and their help on countless occasions. I have enjoyed the pleasant atmosphere in our group owing to its nice (past and present) members: Stefan Wünsch, Juliane Raasch, Alexander Schmid, Philipp Trojan, Matthias Arndt.

I thank the technical and administrative staff for their help during my time at IMS. To Bachelor students, Thomas Silbernagel and Benjamin Bachowsky, I am very grateful for their valuable work and the great time.

Lastly, I would like to thank my family for all their love and encouragement.

Karlsruhe, in November 2017 *Ilya Charaev*

1 Introduction: scientific motivation and aim of this work

Das Lichtquant (light quantum) was the first name of a photon given by Albert Einstein in 1905. The elementary excitation of a single mode of the quantized electromagnetic field have been intensively investigated over last century. The quantum nature of the light provides an unprecedented capability in areas that range from the traditional applications. However, it implies that intensity of incoming radiation will be at quantum limit. The demonstration of possibility of detection of individual photons is of great importance for a variety of scientific and technical applications. In modern times, many research projects require ultra-fast and high sensitive detectors of single photons. These include photonic- and quantum-integrated circuits, spectroscopic applications, astronomy, quantum key distribution, medical diagnostics, remote sensing, deep space communication links for data transmission on interplanetary missions, optical quantum computing and many others [1].

The detectors of single photon are classified by material systems used for their fabrication: photomultipliers, families of semiconductors and superconductors with nearly innumerable variations. The photomultiplier tube is the first single-photon detector which was invented in the thirties of last century. Nowadays, these detectors have a pretty low detection efficiency *DE* in infrared range with huge rate of dark counts (*DCR*).

A single-photon avalanche photodiodes (SPADs) is the strongest candidate for detection of single-photon in near- and mid-infrared spectrum. These detectors demonstrate a high performance in near-infrared range [2] at temperature about 200 K. The shortest timing jitter (FWHM) shown by SPADs is 52 ps [3]. This makes avalanche photodiodes attractive for scientific and industrial applications. However, the spectrum of Geiger-mode avalanche photodiodes is limited by a cut-off wavelength of around 1670 nm. Another issue is the device homogeneity, including the reproducibility and uniformity of device performance [4]. Some semiconductor single-photon detectors can operate at the temperature of liquid helium 4.2 K, 80% detection efficiency has been achieved with InGaAs charge integration photon detectors (CIPD) at 1310 wavelength. The drawback of CIPDs is their strongly limited maximum count rate. Detailed description of recent emerging techniques and future perspectives for SPADs can be found in [5], [6].

As an alternative, the concept of a superconducting nanowire single-photon detector (SNSPD) was demonstrated in 2001. The SNSPD is a promising technology which is offering high sensitivity of single photons combined with low dark count rate, short jitter and recovery time. The detectors operate at a temperature below a critical value, T_C, and are applicable for the use of a He^4 transport Dewar in many cases. First SNSPDs were made from thin polycrystalline NbN [7] and NbTiN [8] films. These materials can be reliably elaborated and patterned; they have a hard superconducting gap and critical temperatures well above the temperature of liquid helium that facilitates their use. SNSPDs show a detection efficiency around 80% at telecommunication wavelength. The *DE* can be further improved by radiation coupling techniques to enhance the absorbance in thin films. As compared to nitrides uniform thin films of amorphous superconductors like WSi [9], NbSi [10], MoGe [11] and MoSi [12] were shown to be more promising for an effective detection of near infrared photons with larger wavelengths. These detectors are sensitive to low-energy photon in range up to 5 μm [13] and have 93% detection efficiency at 1550 nm [9]. The drawback is a noticeably lower critical temperature and correspondingly smaller energy gaps, which unavoidably results in lower critical current densities. Moreover, it was found [11], [12] that SNSPDs from these low-temperature materials exhibit larger timing jitter than those from nitrides [14]. That is why high-quality ultrathin nitride films remain a reference in the highly competitive field of SNSPDs.

In recent years, new concepts of single-photon detection have been proposed. Some of promising single-photon detection technologies are graphene-based detectors [15], carbon nanotubes-based radiation detectors [FKG+13] and modified multi-pixel radio-frequency SNSPD arrays [DKW+17], [DKW++17].

In spite of the wide distribution of SNSPDs, fundamental questions of physical processes of photon detection mechanism are still under discussions. It was accepted that to induce a voltage response from nanowire, sufficient amount of quasiparticles has to be generated after photon absorption. Depending on the energy of the incident photon, the number of quasiparticles changes along with the detection efficiency of the SNSPD. The detection efficiency reaches the highest value when the energy of photon is enough to create sufficient number of quasiparticles. At a certain threshold, the *DE* starts to decrease exponentially towards the infrared range. The restricted spectral bandwidth of SNSPDs limits performance at near-infrared and infrared ranges. The *DE* threshold of SNSPD (so-called cut-off wavelength) was described in frame of analytical and numerical models for SNSPD. The crucial parameter in these models is the critical current I_C. The critical current is usually smaller or even much smaller than the depairing current. The spectral bandwidth of SNSPD could be extended to larger wavelengths via pushing I_C towards the depairing current limit $I_C = I_C^d$. Once the current ratio I_C/I_C^d is enhanced, it would enable operation of detectors (made of materials with high T_C) at relatively high temperatures and high bias currents. The main aim of this work is broadening the spectral bandwidth towards infrared range by increasing the intrinsic threshold wavelength. At the same time, the experimental threshold dependence predicted differently by theoretical models has to be demonstrated.

Commonly used detection models will be introduced and analyzed according to their predictions in view of the cut-off wavelength in chapter 2. The state of the art and theoretical background are reported in the following section. The main limiting factor of the critical current and approaches to extend the spectral bandwidth of SNSPDs are presented in the last section of this chapter.

The ratio of I_C/I_C^d is affected by the following influences: stoichiometry of films, 3D proximity effect and current crowding in the nanowire. The influence of the stoichiometry on performance metrics of detector was systematically studied in [16]. The 3D proximity effect is reduced by improving the quality of patterning and deposition techniques.

The first task is to develop the technology for enhancement of superconductivity in patterned nanowires. The NbN is widely by used as a material for SNSPDs devices by many research groups. The superconducting properties of films can be improved by using a bilayer system with insulating layer. To reduce the roughness of edges in patterned nanowire, the negative electron-beam lithography process is developed with positive resist (chapter 3).

The current crowding effect is negligible when the layout of the structure doesn't have any bends. To avoid current crowding in SNSPD, the multi-layer technology of bend-free SNSPD is developed with a new single-spiral layout. Additionally, the one symmetry bended spiral is specially designed to investigate the detection mechanisms of the single-photon regime in the magnetic field (chapter 3).

The 3D proximity effect in nanowires is described as the lateral NSN proximity system in which the superconducting central band is sandwiched between damaged edge-bands with suppressed superconductivity. The width of the superconducting band is experimentally determined from the value of magnetic field at which the band transits from the Meissner state to the static vortex state. A systematic experimental study of 5 nm thick NbN strips with widths in the interval from 50 nm to 20 μm are presented in chapter 4.

The detection probability is studied for low energy photons to be detected on SNSPDs with different layouts. For samples with each layout the spectra of the single-photon response are measured in the wavelength range from 400 nm to 1600 nm and defined the cut-off wavelength beyond which the response rolls off. An extended spectral bandwidth will be demonstrated on improved SNSPDs (chapter 5).

Using structures in the form of a square spiral, where all bends have the same symmetry with respect to the directions of the bias current and external magnetic field, areas around the bends are studied in terms of the contribution to the rate of photon and dark counts at small photon energies. Comparison of the computed local absorption probabilities for photons and the simulated local threshold detection current is performed to explain asymmetries of performance metrics in magnetic field (chapter 6).

2 Superconducting nanowire single-photon detector

This chapter describes theoretical concepts of single-photon detection by superconducting nanowires. With an overview of the state-of-the-art in SNSPD research the comprehensive report of existing theoretical models is presented. After a brief historical overview, the principle of single-photon detection will be described in details. Achievements over the last decade will be discussed in terms of identification of fundamental problems which are still remaining in understanding of the detection process of single photons. In section 2.2.1, the theoretical models of detection are introduced in particular to describe energy-current relation which is predicted differently in the framework of existing models. Understanding of theoretical basis helps to formulate tasks for experimental work which should prove the key role of the experimental critical current in the position of the spectral cut-off and appropriate dependence of current-energy relation in framework of existed theoretical models. The main mechanisms of limitation of the critical current will be discussed in the next section. Finally, requirements for detectors and approaches for improving the spectral bandwidth of SNSPD are formulated in the last section of this chapter.

2.1 Operation principle and metrics

One of the first experimental observation of detection of individual particles by superconducting strip was reported in 1949 by Andrews et al [17]. The theoretical explanations have appeared much latter in [18] and [19]. It had been realized that particles create a normal spot in a current-carrying superconducting strip and that this spot diverts the current into the still superconducting part of the strip cross-section. The dynamics of the normal spot [18] and the propagation of the normal zone in the strip [20] were described via combination of heat diffusion in the strip and heat transfer to the surrounding. It was soon realized, that the reduction of cross-section of the strip to produce measurable change in the resistance under illumination with optical photon.

In 2001, Goltsman *et al* [21] demonstrated a new superconducting detector concept based on superconducting nanostrips. This type of device, known as the superconducting nanowire single-photon detector (SNSPD), has a single-photon sensitivity in a range from visible to infrared wavelengths. The new concept was based on formation of a hot spot in NbN superconducting film nanofabricated into narrow strip in form of meander (Fig. 2.1).

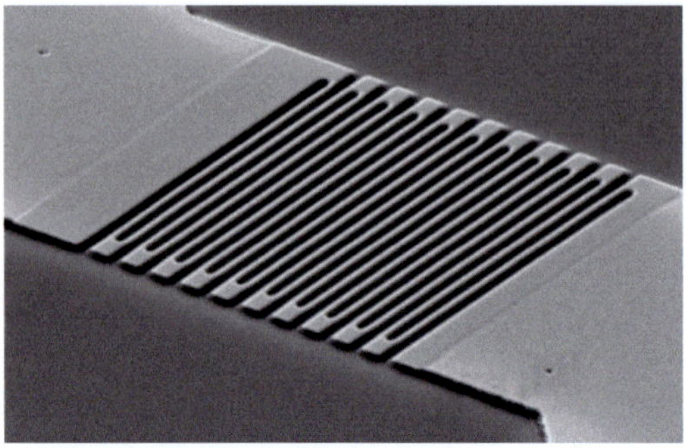

Figure 2.1: The initial layout of superconducting nanowire single-photon detector.

Due to diffraction blur (so-called Airy disk) which is proportional to the wavelength, the diameter of the central disk of light spot is limited by diffraction effect as $2.44\lambda >> W$ (λ – wavelength, W – width of nanowire) [22]. The meander layout allows to reduce significant-ly optical losses. The meander detector is biased close to the critical current I_C at ambient temperature $T << T_C$, T_C - critical temperature.

The initial theoretical explanation was based on the normal hot spot created by an absorbed photon (Fig. 2.2) [7]. After hot spot formation the supercurrent is expelled from the normal region and flow along the periphery of the hot spot. This leads to an increase of density of the supercurrent at sidewalks higher than the critical current density, j_C, and whole area transits to the resistive state. The measured voltage pulse is associated with detection of a single photon and is called pulse-count event. The theoretical basis of the first hot spot theory will be presented in details in section 2.2.1.

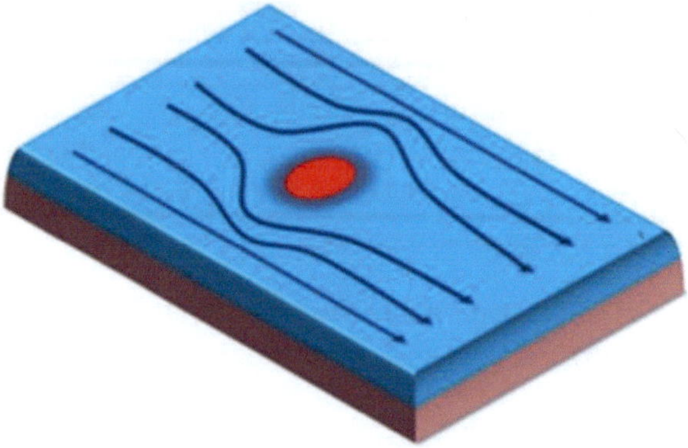

Figure 2.2: The first theoretical model of SNSPD detection. The absorbed photon creates a hot spot in the superconducting nanowire biased at current close to the critical value. The supercurrent is expelled from the hot spot region into the sidewalks. The current is then exceeded in the whole cross-sectional area of the nanowire and a measurable voltage pulse can be read out along the nanowire.

Further intensive investigations over last sixteen years revealed that the mechanism of single-photon detection in a superconducting nanowire is more complicated and vortices are essential for the formation of the initial normal domain that triggers a detection event (see section 2.2.1).

On the phenomenological level, the process of single-photon detection can be split into several stages (Fig. 2.3). In the beginning, the photon should be absorbed by superconducting nanowire. The photon energy in optical and near-infrared ranges is about three orders of magnitudes larger than the superconducting energy gap, Δ, of NbN. Depending on composition of polycrystalline NbN films, their thickness and corresponding square resistance, the absorbance is variated in absolute value and in the spectral range of visible and near-infrared light. The dependence of absorbance on wavelength becomes rather weak in infrared range. Nanofabrication of the film into meander nanowires produces the effect of polarization dependence of absorbance. The absorption probability of photons is much larger when the electric field in the wave is collinear to the nanowire [23]. Furthermore, the local probability of detection strongly variates in a nanowire superconducting single-photon detector [24].

After photon absorption by a Cooper pair, electrons which received energy from the photon become non-equilibrium and they diffuse out of the formed hot spot. If absorbed energy is enough to destroy superconductivity than the core of hot spot is normal. Otherwise, the hot spot remains in superconducting state with partially suppressed superconductivity. Nevertheless, the current flows around the hot spot region. The increased density of the current in sidewalks may be enough for appearance of the resistive state.

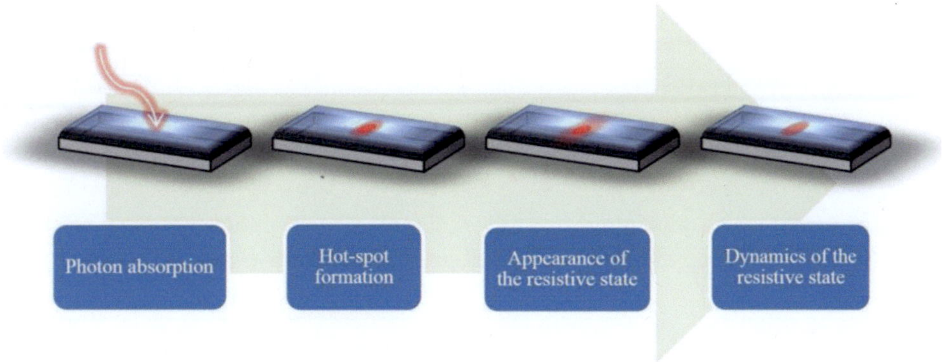

Figure 2.3: The process of single-photon detection including several stages.

The electro-thermal dynamics approach for describing the boundary of the normal domain incorporates several dynamic processes such as heating due to transport current, heat transferring at normal and superconducting sides of boundary. The steady-state and time dependent of the normal domain were described in [25], [26], [27]. The voltage pulse is generated when the normal domain in the cross-section of nanowire grows due to dissipation energy of the current. When the absorbed energy dissipates from the normal region, growing of this zone stops and then it disappears.

Other details of process of emergence and disappearance of the hot spot which are described differently in frame of existing models will be considered in in section 2.2.1.

Metrics

The most crucial performance metric of SNSPD is probability of the count event after incidence of a photon on detector. This metric, which is called detection efficiency *DE* depends on optical properties of the film, detector geometry and energy of incident photons. Together with the quality of the optical coupling in experimental setup, the detection efficiency defines the general characteristic of any setup named a system detection efficiency, *SDE*. Basically, the *SDE* is a ratio of the number of events, detected by the measurement system to the total number of photons delivered to the system during fixed time period. Thus, the system detection efficiency can be presented as a product of three general efficiencies:

$$SDE = ABS \times IDE \times OCE \quad , \tag{2.1}$$

where *ABS* is absorption efficiency which determines the probability of absorption of the incident photons into detector area by nanowires. The absorption depends on the cross-section of nanowire with gap in between two nanowires, absorption coefficient of superconductor and optical path length of the photon; the *IDE* is the intrinsic detection efficiency of the detector itself. The *IDE* can be presented as a ratio of number of output pulses from SNSPD divided by the total number of photons which absorbed by detector. The combination of absorption efficiency and intrinsic detection efficiency is the detection efficiency ($DE = ABS \times IDE$) which was introduced above; the *OCE* is the optical coupling efficiency which describes coupling losses in the optical path to the detector.

The detection efficiency variates with energy of incident photons (Fig. 2.4). It was also found that *DE* changes with changing of the transport current through the detector. For each particular current, there is a cut-off wavelength, λ_C, in the detection efficiency. For wavelengths smaller than λ_C, *DE* is almost constant (filled area in Fig. 2.4) while beyond the cut-off, $\lambda > \lambda_C$, it drops rapidly with increasing wavelength. The pulse from SNSPD can be recorded even without incident photons into the area of the detector.

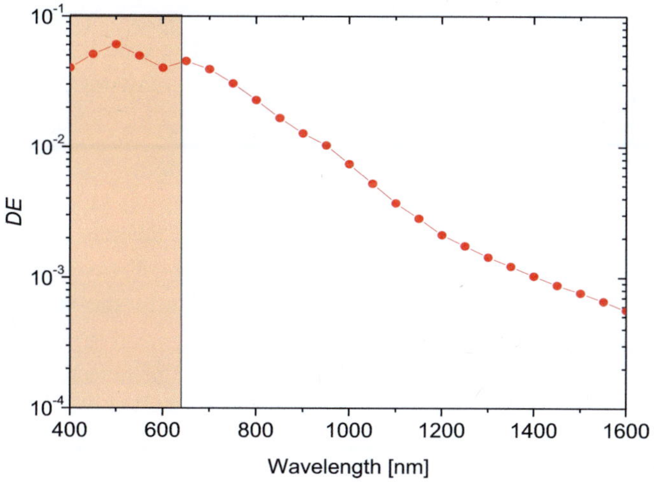

Figure 2.4: The spectral detection efficiency of SNSPD measured in the spectral range between 400 and 1600 nm wavelength at the fixed bias current.

This equivalent of the noise is called the dark count rate, *DCR*, and it appears at the bias current, close to the critical value. The dark count rate is independent of noise level of the readout circuits and could be partially associated with interruption of uniformity of the current in bends of meander SNSPD. The minimum detectable photon flux is limited by rate of the dark count.

Low timing jitter is one of the SNSPDs metrics which is important for many applications and may a have great potential for extension of the application field of SNSPD. The random time delay causes timing jitter in the appearance of a voltage transient, which a photon detector generates. The measure of the timing jitter is the width of the statistical distribution in the arrival times of voltage transients with respect to the corresponding photon absorption times. In SNSPD applications, timing jitter limits the accuracy of measurements of photon arrival times.

2.2 State of the art

During the last 15 years the technology of superconducting nanowire single-photon detectors was under intensive development, continuously improving performance of the SNSPD. Many efforts in the fields of optics, solid state physics and thin-film technology have been attempted in order to find the way to increase detection efficiency (*DE*) at longer wavelengths and to decrease timing jitter and dark count rate (*DCR*) of detectors. The development of SNSPDs follows several main directions. The search for optimal materials and the improvement of the quality of superconducting nanowires are among the most straightforward ones.

Spectral detection efficiency

Several experiments in investigations of spectral detection efficiency have shown a current dependent threshold in the detection efficiency [28]. The cut-off wavelength increases with the increase of the current. The photon count rate (*PCR*) saturates when the current approaches the critical current but decreases rapidly when the current decreases below the threshold value, I_{det}, which is called detection current [29]. The detection efficiency is limited by absorbance of the given film. A great improvement in absorbance was achieved by patterning a double-cavity and a mirror at 1550-nm wavelength [30]. This makes the fabrication process more complicated.

The decrease of the cross-section of nanowire (width and thickness) results in shift of the cut-off wavelength towards infrared range [31], [32]. Beside the geometry, the cut-off wavelength depends on the operation temperature [33]. For temperatures well below the transition temperature, λ_C increases with the decrease of the operation temperature.

The shape of transition from the hot spot response spectrum with almost constant of *DE*, to the range of quickly dropping the efficiency, depends on operation temperature, properties of superconducting materials and quality of fabricated nanowires. The broadening of transition region between constant and fast decreasing dark count rate is associated with the presence of bends in a layout and constrictions in nanowires. Furthermore, multiphoton events may dominate photon count rate and broaden the transition. The absorption of several photons is required to the multiphoton event within the life-time of the hot-spot at a distance from each other less than the diameter of the hot-spot. The contribution of multiphoton events to photon counts cannot be excluded absolutely. However, it was proved that multiphoton events don't play crucial role in the position of the cut-off wavelength [34].

All hot-spot models of the detection mechanism predict larger cut-off wavelength for the superconductors with smaller energy gap Δ. Indeed, recent experiments with TaN [35] and WSi [9] meanders have demonstrated larger λ_C, than it was found for meanders from widely used NbN films.

Although it has been commonly agreed that λ_C increases with the bias current I_b, the exact relationship between I_b and λ_C remains controversial [28]. Since the relative current I_b/I_C^d cannot be larger that I_C/I_C^d, where I_C is the experimentally achievable critical current, the way to push λ_C further towards larger values is to increase I_C. Despite of the advanced technology, in practical devices I_C remains smaller than I_C^d (~50%) [28]. In NbN meanders, the ratio I_C/I_C^d can be increased to some extent by variation of the film stoichiometry [16].

Recently, a double-spiral layout was suggested as a promising alternative for design of single-photon detectors [36]. The absorption of spiral detectors is independent of polarization [37]. In comparison to meanders, detectors with the double-spiral layout demonstrated larger I_C, λ_C and detection efficiency. However, the cut-off wavelength was found only slightly higher than λ_C of meander detectors (~600 nm) measured at 4.2 K.

Different materials

Beside widely used crystalline NbN, many other superconducting materials were chosen as alternative material for SNSPDs. A particular interest exists in materials with a smaller superconducting energy gap, Δ, than in NbN. Naturally, the first material of the initial superconductor for SNSPDs was pure Nb with bulk T_C = 9.25 K, which has the additional benefit that nanowire detectors made from Nb have shown improved dead times compared to NbN [38]. However, the cut-off wavelength of Nb SNSPD is limited by the large diffusion coefficient. In this case, the size of hot spot is pretty large when electrons are thermalized which leads to locally smaller heating and weaker influence on superconducting properties.

The observed increase of detection efficiency of SNSPDs based on TaN film with T_C ~ 10 K at longer wavelength was reported in [35]. However, significant improvement in the detection efficiency was obtained on SNSPDs made of amorphous WSi film. As compared to nitrides, uniform thin films of amorphous superconductors like WSi, NbSi, MoGe and MoSi were shown to be more promising for an effective detection of near infrared photons with larger wavelengths. Because the coupling energy of Cooper pairs decreases with the transition temperature, the same number of non-equilibrium electrons is excited by photons with lower energy. The drawback is a noticeably lower critical temperature and correspondently smaller energy gap, which unavoidably results in lower critical current densities. Furthermore, the maximum of the count rate and minimum timing jitter are limited by rather long relaxation time. It was found that the life time of the hot spot in amorphous WSi film is more than one order of magnitude longer (τ~500 ps) [ZEW+16] than in polycrystalline NbN film (15 ps) [39]. This difference was observed in absolute value of the life time as well as its current dependences in NbN and WSi materials. The life time grows, significantly, in WSi with applied current, while it's almost independent of the current in NbN. The difference in current dependence is associated with the size of the hot spot which is formed in the nanowire after absorption of the photon. The size of the hot spot in WSi nanowire was found to be larger than the width, while the hot spot in NbN occupies only a fraction of the nanowire [ZEW+16].

Beside amorphous materials, superconductors with similar superconducting energy gap than in NbN have been tested as candidates for SNSPDs. NbTiN nanowire devices have been shown to possess low kinetic inductance due to their low resistivity above the superconducting transition temperature. The single-photon sensitivity of nanowires made of MgB_2 with the critical temperature 35 K was demonstrated at visible and infrared ranges [40]. However, the quality of the films has to be optimized in order to fabricate structures with less than 100nm feature size. The perspective of high-temperature superconductors as a material for detector of infrared photons is still under discussion. Due to the low ratio between specific heat capacities of electrons and phonons in HTS nanowires, the detection of near-infrared photon is possible only at currents close to depairing current, I_C^d [41].

Dark Count Rate

Statistical fluctuations in the form of dark counts restrict the minimum detectable photon flux. At the inner edges of bends and turns, where the supercurrent flow around a sharp corner, the local current density increases, which causes a local reduction of the energy barrier for nucleation of magnetic vortices. Among different topological fluctuations, hopping of vortices across the strip is commonly considered as a mechanism of dark counts. Hence, bends and turns with sharp corners are places from which dark counts most probably originate. Only a few indirect experimental verifications of this suggestion have been reported. Engel *et al.* [42] found in their meanders a slight asymmetry in the rate of dark counts with respect to the magnetic field direction and assigned it to differences in shapes of right and left turns. Akhlaghi *et al.* [43] showed that, in a nanowire with a single bend, rounding the sharp inner corner of the bend results in an increase of the critical current and in a reduction of the dark count rate (*DCR*) of the whole structure. Lusche *et al.* [44] found differences in current dependencies of the vortex energy barrier in the case of photon and dark counts and associated them with different locations of these events. This can be either a single vortex near the strip edge or a vortex-antivortex pair (VAP) close to the midline of the strip. Vortices are then swept by the Lorentz force across the strip. The energy dissipated along the trajectory of the vortex in the strip causes the formation of a normal belt. In the fluctuation-assisted scenario, a vortex crosses the nanowire with certain thermodynamic probability through the segment, where the energy barrier is reduced due to photon absorption.

Timing jitter

During last decade many groups have reported a small timing jitter in SNSPDs. However, measurements were done with different detector layouts and electronics that hamper direct comparison. A system jitter with full width at half maximum (FWHM) as low as 18 ps has been demonstrated for SNSPD based on NbN [45]. It has been realized that electronic noise severely enhances system jitter and introduces the current dependence of the jitter. Subtracting the noise contribution, one obtains jitter inherent to the detector itself which is called intrinsic jitter. It was found that intrinsic jitter increases in nanowires with smaller thickness and larger kinetic inductance per unit length [46]. Furthermore, jitter increases with the size of the detector [45] and is less for the central part of the detector area as compared to peripherals [47]. Although low jitter itself is a challenge, it becomes attractive only in conjunction with the practical values of the detection efficiency and maximum count rate. Since the size of the detector affects differently these two metrics, the size stays necessarily in the list of trade-off parameters. Jitter increases in nanowires from superconducting films with low transition temperatures. For nanowires from WSi [9] and MoSi [12], jitter is almost one order of magnitude larger than in nanowires from NbN.

While instrumental aspects of the system timing jitter have been thoroughly discussed, physical mechanisms of the intrinsic jitter remain largely unclear. Revealing those mechanisms should give the answer how to decrease jitter value and what is the limit. For WSi, Fano fluctuations were shown to broaden the decay of the photon detection efficiency with the decrease of the current trough the nanowire [48]. They should also affect the time delay between photon absorption and the emergence of the resistive state. Whether this mechanism may affect jitter in NbN nanowires is not clear. The delay of the resistive stay in NbN was found to exceed 65 ps [49]. It has been shown that this delay may depend on the film thickness via the escape of non-thermal phonons to the substrate [50]. Recently, spread in the traveling time of a magnetic vortex across the nanowire was analyzed [51]. For NbN, it can contribute a few picoseconds to the intrinsic timing jitter.

2.2.1 Phenomenological theories and detection models of SNSPD

In this section, commonly used detection models will be introduced and compared in terms of their predictions and limits.

Normal core hot-spot model

As was mentioned in section 2.1, the energy of incident photons is much larger than the superconducting energy gap of low-temperature superconductor. An absorbed photon in the nanowire excites an electron with high energy. Due to subsequent thermalization of exited electron, quasiparticles are generated in a cascading process. After thermalization time, the number of quasiparticles, n_{qp} can be approximately estimated as:

$$n_{qp} \approx \frac{h\nu}{\Delta}, \tag{2.2}$$

where h is the Planck constant and v is a frequency of the photon.

The quasiparticles diffuse in a cloud of growing diameter. The normal core can be formed in the center of a cloud if superconductivity is locally suppressed. In order to redistribute the transport current in nanowire, diameter of formed normal-conducting core should be not smaller than coherent length, ξ. Then the transport current flows around core with suppressed superconductivity (Fig. 2.5). This should provide an increase of current density into sidewalks. When the diameter of the hot-spot is big enough, the superconductivity is suppressed through the cross-section of the nanowire by reaching the critical value of the current density in the sidewalks.

The normal-core hot-spot model predicts a square-root dependence of current-energy relation for 2D system as:

$$I_{\text{det}} = I_C\left(1-\gamma\sqrt{E}\right), \tag{2.3}$$

where γ is the material and device dependent proportionality factor; E is the energy of the absorbed photon. It has to be noted that γ is different for each model.

The normal-core model predicts a dependence of the cut-off wavelength on bias current:

$$\lambda_c = \frac{32 e^2 R_S D h c \varsigma}{\pi W^2 \Delta^2}\left(1-\frac{I_b}{I_C^d}\right)^{-2}, \tag{2.4}$$

where c is the speed of the light; ς is conversion efficiency of the photon energy from initial excitation in the film to quasiparticles at the superconducting gap (fit parameter); W is the width of nanowires; D is the quasiparticle diffusion coefficient; R_S is a square resistance.

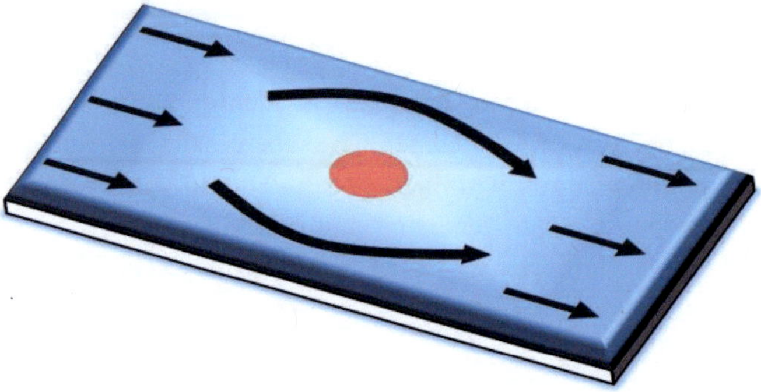

Figure 2.5: A sketch of the normal-core hot-spot model to describe the detection mechanism of SNSPDs.

The numerical normal-core hot-spot model predicts a nonlinear behavior outside the experimentally measured range. Moreover, the position dependence of the detection threshold cannot be presented in framework of this model.

Diffusion hot-spot model

The refined hot-spot model appeared several years later after publishing the first theoretical approach for detection of single-photons by current carrying superconducting nanowires. Instead of the normal core of the hot spot, the number of Cooper pairs within a section with dimensions about coherent length was considered. Although the number of such Cooper pairs is reduced due to photon absorption, they still carry the current with acceleration. Once the critical velocity of Cooper pairs is achieved, they break up with following transition of nanowire into the resistive state (Fig. 2.6). Since the current carrying capacity is correlated with total number of remaining Cooper pairs, the expression of current-energy relation presented as:

$$I_{\text{det}} = I_C\left(1-\gamma E\right), \tag{2.5}$$

The diffusion hot-spot model describes analytically the cut-off of hot spot response spectrum. The energy of the photon which is needed to detect uniquely absorbed photon depends on geometrical parameters and material properties of superconducting nanowires as well as the ratio of the critical current to depairing critical current:

$$\lambda_C^{-1} = \frac{3\sqrt{\pi}}{4hc\varsigma}\Delta^2 WdN_0\sqrt{D\tau}\left(1-\frac{I_C}{I_C^d}\right), \tag{2.6}$$

where d is the thickness of the superconducting film; N_0 is the density of electron states; τ is the relaxation time for a given superconductor.

Although the expected linear dependence of the threshold on width and thickness as well as corresponding dependences on material properties were experimentally confirmed, the expected temperature down-shift of the threshold is not in agreement with experimental results obtained at relative temperature $T/T_C \geq 0.5$. Moreover, the explanation of the position dependence of the threshold is impossible in the framework of the diffusion hot-spot model.

The weakness of both normal-core and diffusion hot-spot model is the prediction of only deterministic, threshold-like responses. The exponentially decrease of count rate with reduction of the bias current is associated with some thermal or quantum fluctuations which are not taken into account in the described above models.

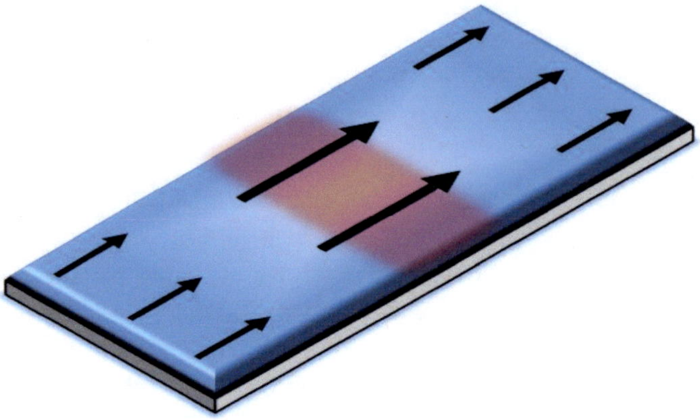

Figure 2.6: A sketch of the diffusion-based hot spot model to describe the detection mechanism of SNSPDs.

Photon-triggered vortex-entry model

The vortex-antivortex pair (VAP) is one sort of fluctuations which was taken into account in the following vortex-assisted models. It was assumed that VAP is responsible for the slow drop of the detection efficiency and lower current than I_{det}. Soon, single-vortex hopping was suggested as a mechanism of single-photon detection for the range beyond the threshold (so-called probabilistic regime) (non-filled area in Fig. 2.4) [32] and later for whole response spectral range (non-filled and peach areas in Fig. 2.4) [52]. In the framework of the model, the vortex penetrates at the place with a reduced barrier for vortices due to photon absorption (Fig. 2.7). Based on the London model and theoretical investigations, the analytical expressions of the dependence for the detection efficiency on temperature, current and magnetic field, as well as dark count rate are given in the photon-triggered vortex-entry model.

The current-energy relation is predicted in the framework of this model as:

$$\frac{I}{I_{c,\nu}} = \exp\left(\frac{C}{\nu-(E+1)}\right), \qquad (2.7)$$

where $I_{\mathrm{c,v}}$ is the critical current that reduces the energy barrier for vortex penetration to zero; C is a constant; ν is the vortex energy scale in the initial state. Obviously, the behavior of the current-energy relation is highly non-linear.

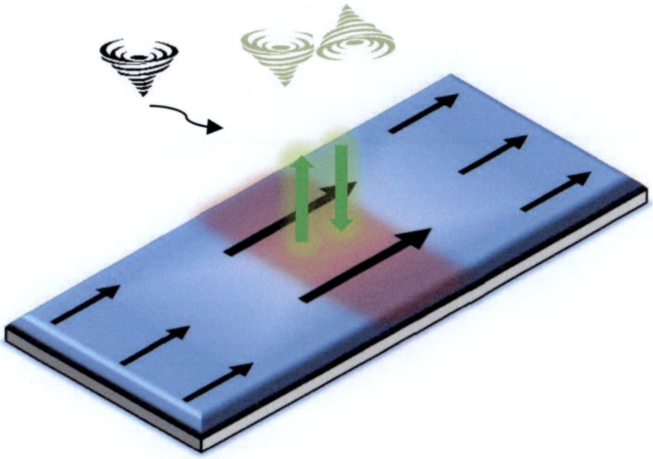

Figure 2.7: A sketch of the photon-triggered vortex-entry model to describe the detection mechanism of SNSPDs.

Normal-core vortex model

The concept of the normal-core vortex model is based on the assumption, that photon absorption results in a local increase of the effective temperature in the area with fixed radius (similar "normal core" detection model). The reduction of superconductivity in the nanowire after photon absorption and subsequent formation of the resistive state are described by the time-dependent Ginzburg-Landau equation, heat-diffusion and Poisson's equations. However, the time-dependent Ginzburg-Landau (GL) equation is appropriate for cases when the temperature is close to the critical value. Before this issue was resolved, several experimental results were not successfully described by the scenario of vortex penetration to the normal core.

For calculations, a fixed radius was used for the area of increased effective temperature. In order to describe the generation of quasiparticles, the nonequilibrium distribution function was used with variated parameter inside and outside of hot spot. The relaxation method was used to find stationary solutions of the GL equation.

Based on the obtained results, two situations can be considered as detection event depending on the position of photon absorption. In the first case, the photon is absorbed near the edge of nanowire (Fig. 2.8). The local increase of the current at the edge of the (so-called current crowding; it will be considered in the next section) favours the penetration of a vortex in the nanowire. The single vortex entries to the nanowire at some minimum current which is required for penetration of the vortex from the edge to the hot spot.

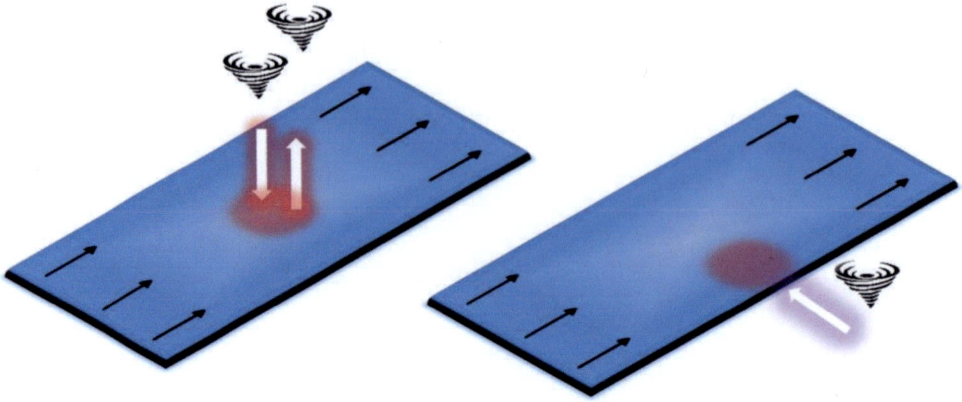

Figure 2.8: A setch of the normal-core vortex model to describe the detection mechanism of SNSPDs. On the left image, the vortex-antivortex pair is generated inside the area in the center with fixed radius. On the right image, a single vortex penetrates into the superconducting nanowire at the edge close the place of the absorbed photon.

A slightly higher current is needed that the vortex overcomes the pinning force and move further across the nanowire. Thus, the motion of a vortex across nanowire results in a voltage pulse. In another case, the photon is absorbed far away from the edge of nanowire. A certain current is required for formation of a vortex-antivortex pair inside the hot spot. Similar to the previous situation, a slightly higher current is needed to overcome the binding energy of the VAP.

In the framework of this model, the threshold current can be calculated as a function of the photon energy. The model predicts a non-linear energy-current relation for low energies of photons. The nonlinear behavior is associated with realistic description of the vortex core in GL model. Unfortunately, the current-energy relation has so far not been observed for low energies (about 0.5 eV).

The normal-core vortex model allows to describe the qualitative position dependence of the detection current, detection efficiency and dark count rate obtained in the magnetic field.

Quantum detector tomography

The method of quantum detector tomography (QDT) is not considered as comprehensive theoretical model of single-photon detection. The current-energy relation $I_{\mathrm{det}}(E)$ cannot be estimated as in the described detection models above. This method characterizes the response of the detector to different numbers of photons at low probability (1% in [53]) of detection event. Other details regarding the quantum detector tomography can be found in [54]. It has to be noted, that all measurements have been done on, so-called, nanodetectors with equal length and width in the range ~150 nm, while widely used meander layout has a length about three orders of magnitude longer. The QDT predicts the almost constant value of the linear efficiency in probabilistic regime which is incompatible with Eq. 2.7. Furthermore, the measured response published in [55] for NbN detector is in contradiction to all data obtained with QDT. In the same publication, the energy–current relation determined for WSi obtained from data published in [29] contains errors and can also not be seen as an indication for a nonlinear energy–current relation.

Generally, the current-energy relation for conditions described above can be written from QDT as:

$$I = I_0 - \gamma E\,, \tag{2.8}$$

where $I_0 \approx 0.75 I_C$.

Comparison of current-energy relations from various detection models

The presented, commonly used, detection models predict different current-energy relation for detection of single photons (Fig. 2.9). According to the normal-core hot-spot model $1\text{-}I_b/I_C^d \propto \lambda_C^{-1/2}$, while the diffusion hot-spot model predicts a linear relationship $1\text{-}I_b/I_C^d \propto \lambda_C^{-1}$, where I_C^d is the depairing critical current. The numerical model invoking Ginsburg-Landau approach to the evolution of the order parameter results in a $1\text{-}I_b/I_C^d \propto f(\lambda_c)$ dependence which falls between the two analytical relations above.

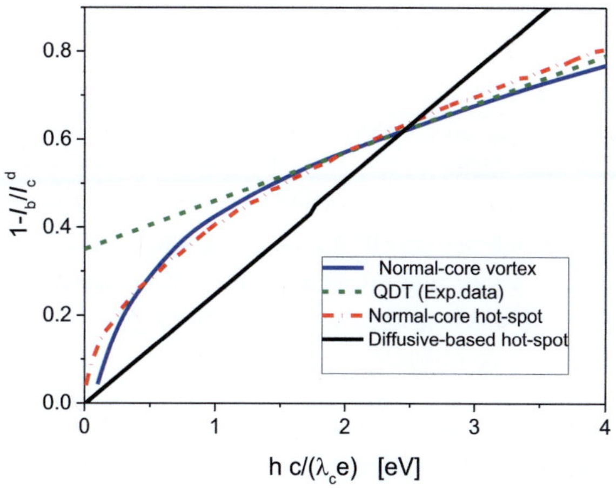

Figure 2.9: Inverted relative current $1\text{-}I_b/I_C^d$ versus energy of photons at the cut-off wavelength. Lines show the current-energy relation obtained with different theoretical models. The models are specified in the legend.

Since the relative current I_b/I_C^d cannot be larger that I_C/I_C^d, where I_C is the experimentally achievable critical current, the way to push λ_C further towards larger values is to increase I_C. Despite of the advanced technology, in practical devices I_C remains smaller than I_C^d.

2.2.2 Restriction mechanism of the critical current in nanowires

As shown in the previous section, the useful spectral bandwidth of SNSPDs can be extended to longer wavelengths via pushing I_C towards the depairing current limit $I_C = I_C^d$. Once the current ratio I_C/I_C^d is enhanced, it would enable operation of detectors (made of materials with higher T_C) at relatively large temperatures and high bias currents. Despite extensive efforts, the experimental values of the critical current in submicrometer lines are usually about 50% of the deparing critical current or even less. As a result, the reduced critical current restricts the performance of superconducting devices.

Since it has become clear that the measurable critical current remains noticeably smaller than the depairing critical current I_C^d [56], [57], [58] many observations have been reported on attempts to understand the physics of the reduction. First, the depairing critical current was almost achieved in relatively thin and narrow strips where the current density is distributed evenly across the width and the thickness [59]. Second, a local perturbation in the strip, which provides local decrease of the superconducting order parameter Δ, restricts the measurable critical current in the current-carrying state. Third, it was predicted theoretically [60] and confirmed experimentally [61], [62], [63] that current crowding near bends of the current path in mesoscopic structures reduces the experimental critical current.

The increased local current density in turns reduces locally the barrier height for vortex penetration and defines the critical current of the whole nanowire (Fig. 2.10). According to theory [60], the reduction factor of the critical current due to a bend in the nanowire depends on the angle of the bend, the wire width W, and the radius of the bend r_b. The effect is present in thin nanowires when $W << \Lambda = 2\lambda^2/d$, where Λ is the Pearl length and λ is the magnetic penetration depth. For strips with the same nominal width and different bending angles, the largest I_C was measured in the straight strip (absence of bends). Current crowding could be eliminated completely in spiral structures.

Beside artificial increase of the current density due to bends, geometrical and structural constrictions in nanowires limit the current carrying capability of nanowire. Constrictions together with bends have a twofold impact on detectors. Moreover, the effective cross-section available for current flow is limited by constrictions. The amount and art of constrictions depend on the nanowire material and fabrication technology.

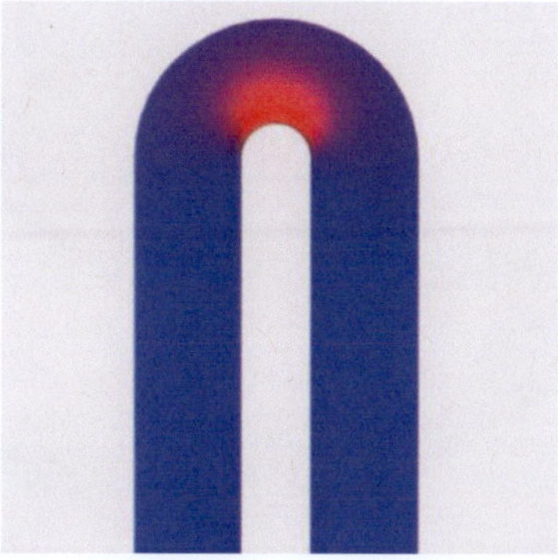

Figure 2.10: Simplified sketch of distribution of the current in a 180° bended nanowire. The local increase of the critical current density at the vicinity of inner corner in bend disrupts the homogeneous flow of the transport current.

Reduction of the film dimension leads to superconducting-insulator transition [64]. In superconducting strips, transition temperature decreases with the decrease in thickness and nominal, geometric width W due to the proximity effect between the normal bands and the superconducting core (Fig. 2.12).

Although the reduction of critical temperature in submicrometer stripes was explained theoretically, the experimental results of $T_C(W)$ dependence have been presented poorly. It was found that, contrary to T_C, the evolution of the density of the critical current as function of the nominal stripe width $j_C(W)$ was non-monotonic. Moreover, the calculated values of j_C are always smaller than the density of the depairing critical current. Only a few studies reported on the stripe-width dependences of T_C and j_C and they are often done on different samples.

Although the reduction of critical temperature in submicrometer stripes was explained theoretically, the experimental results of $T_C(W)$ dependence have been presented poorly. It was found that, contrary to T_C, the evolution of the density of the critical current as function of the nominal stripe width $j_C(W)$ was non-monotonic. Moreover, the calculated values of j_C are always smaller than the density of the depairing critical current. Only a few studies reported on the stripe-width dependences of T_C and j_C and they are often done on different samples.

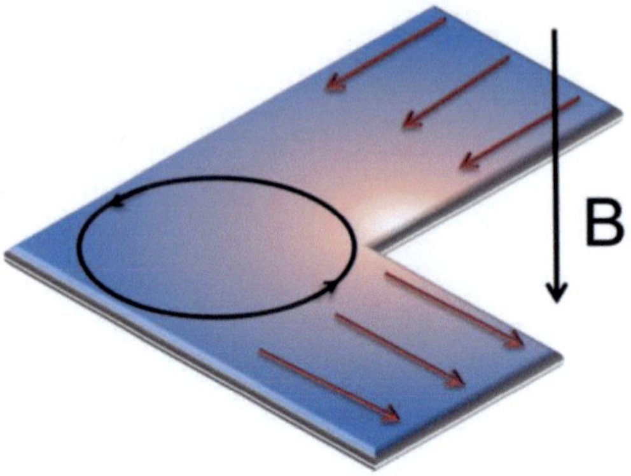

Figure 2.11: The enhancement of current carrying capability of bended nanowires by applying an external magnetic field. The external magnetic field generates a screen current in nanowire which leads to a variation of the current direction depending on field-direction. When the direction of current is different (as shown on the sketch), the maximum applicable current to the nanowire increases due to reduction of current crowding at the inner corner. In opposite situation, the critical current is further suppressed due to locally increased current density in the bend.

Besides that, the influence of the edge quality on superconducting properties of stripes remains unclear. Intuitively, one expects rough edges to cause larger number of defects and correspondingly a reduction in the experimental critical current. The superconducting current circumvents defects. This phenomenon leads to a non-uniform distribution of the current density which peaks at locations adjacent to the defects. Additionally, superconductivity deteriorates at the edges due to mechanical damaging via etching and subsequent oxidation when exposed to air.

There have been already several reports on possible approaches towards enhancement of the I_C/I_C^d ratio. It has been suggested in Ref. [65] and has also been confirmed experimentally [66], [62] that an external magnetic field with the proper orientation diminishes the effect of current crowding (Fig. 2.11). This approach is not suitable for meanders due to alternating turns with opposite symmetry. A special layout with one symmetry of turns has to be designed for experiment in the magnetic field. The practical disadvantage of layouts with alternating straight wires and bends is their strong sensitivity to light polarization. The extinction factor increases with the increase in the wavelength [23].

Recently, a double-spiral layout was proposed as an alternative design for single-photon detectors [36]. Although the performance of photon detectors with the double-spiral geometry becomes better, the sharp S-bend in the current path through the center of the double-spiral layout still limits the experimental critical current of the specimen.

Figure 2.12: The proximity effect in superconducting film. The deposited film of a given thickness d on a substrate has a superconducting core (blue color) and normal-conducting layers d_N (dark blue color). The critical temperature of the whole film is reduced due to proximity effect of the normal layers. The suppression of T_C becomes significant with decrease of nominal thickness. Similar effect was reported for very narrow nanowires made from thin superconducting film [67].

In general, the detection efficiency is higher for SNSPDs with local values of the I_C/I_C^d ratio uniformly distributed over nanowires and bends. As was mentioned, this ratio is affected by different types of defects such as cross-section variations due to the non-uniformity of the thickness or width of the film, nanowire edge defects or internal structural defects weakening the superconducting order. All those effects reduce the current ratio and make characteristics of detectors worse. Therefore, the development and optimization of approaches, which target to a growth of highly uniform thin films and pattering them in defect-free superconducting nanowires, are in permanent focus of numerous research groups. It has been shown that the adjustment of stoichiometry of NbN films towards higher Nb-content results in an increased I_C/I_C^d ratio and thereby in a broadening of the spectral bandwidth of SNSPDs [16]. A large growing rate of thin NbN films increases the granularity of the film while an enrichment with nitrogen increases the amount of amorphous phase but reduces the achievable critical current density.

Another perspective is an improvement of the quality of patterned nanowires. It was found that a straight nanowire made of high-quality films cannot carry the depairing current. The effect arises from partial damage of the nanowire edges during etching. The geometrical uniformity of edges affects the critical current as well.

2.3 Requirements for detectors

Previous achievements described in section 2.2 have shown, that the performance of SNSPDs can be improved to a competitive level for radiation in the optical or near-infrared range. In particular, the spectral bandwidth of SNSPD has been extended by optimization of stoichiometry of NbN film. The shifting of the deposition range towards Nb phase results in increase of the ratio of the experimental critical current to the theoretical limit for given conditions.

As mention above, the experimental critical current is restricted by proximity effect and non-uniformity distribution of the current through the nanowire due to locally increased density of the current in the vicinity of turns and at edges with depleted superconductivity larger than the coherent length ξ. To enhance the ratio of the experimental critical current to depairing current, the proximity effect and current crowding have to be reduced. This can be done by different approaches. First of all, the usage of a bend-free layout for SNSPD such as a single spiral prevents the current crowding which appears in bended structures. The development of a multi-layer technology for single-spiral SNSPDs is required. Second, the proximity effect in superconducting films is reduced by *in-situ* deposition of an AlN/NbN bi-layer. The parameters of sputtering process have to be adjusted. Third, to decrease the roughness of nanowire edges the resolution of electron-beam lithography has to be improved.

The detection efficiency of an optical system based on SNSPDs is limited in particular by the absorption of the nanowire. The overall detection efficiency can never exceed the absorbance unless the light is coupled into the film by other means. Since the material variation changes the electronic state, it is expected that the absorbance of the film will also vary. Hence, the adjustment of stoichiometry is important for device characteristics.

To investigate the role of straight and bended parts of a SNSPD in the optical response, measurements in magnetic fields have to be performed. The energy barrier of detector is lower in the bends than in the straight parts due to the locally increased current density. This also leads to an increase of the vortex-crossing rate in the bends. The vortices are essential for the formation of the initial normal-conducting domain that triggers a detection event. Screening currents imposed by the magnetic field compensate the increase of the current density in the vicinity of the inner corner of a sharp bend in superconducting mesoscopic structures. The standard meander-type SNSPDs demonstrated a fully symmetric behaviour of the critical current in magnetic fields with its maximum at zero field [44]. Because of the opposite symmetry of adjacent turns in the meander design, for the same direction of magnetic field the current crowding effect becomes weaker in one turn but decreases the energy barrier for vortex penetration in the adjacent turn. Hence, a special one-symmetry bended detector and an experimental setup with a superconducting coil are required.

3 Technology of thin NbN films and nanowires for SNSPD

Among superconducting materials niobium nitride is one of the most intensively studied superconductors. The mechanical and chemical stability and rather high critical temperature make NbN attractive for many applications. Great results were obtained on several kinds of detectors based on NbN films, such as superconducting quantum interference devices (SQUIDs) for measuring weak magnetic fields [68], sensitive hot-electron bolometer mixers (HEB) for astrophysics at terahertz frequencies [69], arrays of Josephson junctions for high frequency integrated receivers [70], [PAK+13]. In research area of the SNSPD, NbN films are a widely used material with relatively low quasiparticle coefficient. This allows the formation of the hot-spot in nanowires under near-infrared irradiation. However, patterning of nanowires results in a decrease of the critical temperature. The 3D proximity effect (described in section 2.2.2) in submicrometer structures, the roughness of edges and defects in the films are reasons of a reduced critical temperature and maximum measured critical current in nanowires. The 3D proximity effect can be partially suppressed by an in-situ deposition of a protection layer of isolator as predicted in [71]. Nevertheless, the line-edge roughness impacts on the performance of fabricated devices. Therefore, development and optimization of approaches, which target the growth of highly uniform thin films and pattering them in defect-free superconducting nanowires, are in permanent focus of numerous research groups. Besides that, multilayer processing influences on the superconducting properties of the final device. This may include a patterning of double-cavities and mirror or an integration of SNSPDs into waveguides for improvement of optical absorption. The multilayer technique for a new type of SNSPD in form of a single spiral has to be developed taking into account all aspects of technology described above.

In recent investigations [72], the absorption coefficient itself and its dependence on film thickness, material variation and wavelength of incident photons were studied in details with ellipsometric measurements. It was shown that the absorbance probability decreases with reduction of thickness of the film. Although dependences of absorbance for different thicknesses have the similar trend of increase in spectral range, the absolute value of absorbance correlates with stoichiometric parameters of deposited films.

In the following section, reactive magnetron sputtering of NbN films is described with focus on influence of stoichiometry on superconducting and optical properties of the deposited films. The properties of in-situ sputtered NbN/AlN double layers are compared with films of the same thicknesses without protection.

The optical properties of films deposited at different currents are investigated in terms of applications for detectors. The detection efficiency is limited by the film absorbance unless the light is coupled into the film by other means. Besides estimation of absorbance of films which is important in absolute values, the absorbance of films on different substrates was also obtained from optical measurements of NbN film illuminated from back side of the substrate.

A high-resolution negative electron-beam lithography process based on positive resist which was proposed as an alternative for standard positive lithography is described in details. A comparative analysis for negative- and positive-PMMA processes has been done by edge roughness measurements and comparison of edge profiles. A detailed description of the developed technology for single-spiral SNSPD is presented in the last part of this chapter.

3.1 Reactive magnetron sputtering of thin NbN films

The properties of superconductive and resistive state of NbN almost always correspond to films where the conditions of deposition were optimized to obtain films with maximum T_C. However, the dependence of these properties on thickness and width of nanowires as well as on variations of deposition conditions makes it difficult to compare structures made from different films. Therefore, investigations of material properties of the NbN films is required before patterning them into detectors and single nanowires.

The fabrication process of SNSPDs starts with the deposition of a niobium nitride thin film on a R-plane cut, one-side polished sapphire substrate by reactive magnetron sputtering from a pure Nb target in an atmosphere of argon and nitrogen at a partial pressure of argon $P_{Ar} = 1.9 \times 10^{-3}$ mbar and nitrogen $P_{N2} = 4 \times 10^{-4}$ mbar. The discharge characteristic of the reactive sputtering process for given pressures is presented in Fig. 3.1a. The relevant working points are marked by the gray area in the graph. At low currents <100 mA, the sputter process is slower than the chemical reaction and the deposited film has a higher nitrogen content. At high currents >120 mA, Nb is sputtered from the surface faster than it can react with the nitrogen, causing a discharge characteristic very similar to the one for pure Nb sputtering in argon atmosphere. The detailed description of the deposition process of NbN thin films can be found in [16].

During deposition, the substrate was placed without thermal glue on the surface of a holder which in turn was placed on a heater which was kept at 850°C. Several films at different discharge currents I_{dis} were deposited with a thickness of $d \approx 5$ nm, which was measured by a stylus profilometer. In order to sputter films with the same thickness, the rate of deposition has been estimated for the whole range of current.

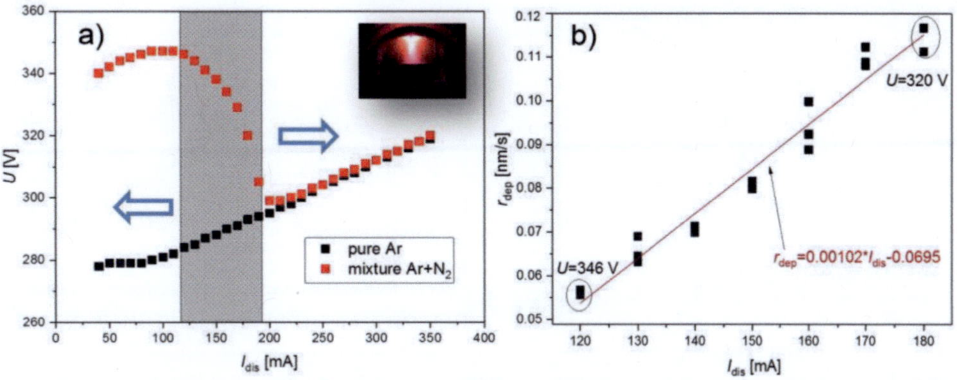

Figure 3.1: a) The current-voltage characteristic (CVC) of reactive magnetron sputtering of NbN. b) The rate of deposition for different currents. Deposition rates of NbN films were evaluated for deposition conditions marked by the gray area in Fig. 3.1a. In the considered range the rate is to a good approximation proportional to discharge current. The respective linear fit is displayed as the red solid line.

The obtained result is plotted on Fig. 3.1b. The rate of deposition is proportional to the discharge current. The voltage U decreases from 346 V down to 320 V at deposition currents covering the gray area in the graph.

3.1.1 Transport properties

The temperature dependence of the film resistance was measured immediately after deposition in the temperature range from 4.2 up to 300 K using a standard four-probe technique. The change in the stoichiometric composition of the films results in a change of T_C. In Fig. 3.2a, it's clearly seen that the critical temperature increases with increasing discharge current. However, the deposited film at the highest current in range of choice has lowest T_C. The wider transition to the superconducting state of film deposited at highest current associates with increased content of Nb which has T_C (9.2 K for bulk material) lower than NbN. The reduction of resistance in all range of I_{dis} supports the assumption that with increasing sputter currents, the Nb content of the film is higher.

The specific resistivity, $\rho = R_{sq} \times d$ is evaluated using the measured thickness and the square resistance of the films. The specific resistivity of the film with highest T_C in the normal state just above the transition ($T = 25$ K) is ρ_{25}=126 μΩ×cm. The residual-resistivity ratio $RRR = \rho(300\text{K})/\rho(25\text{K})$ is found for the film with highest T_C slightly below one, $RRR = 0.98$.

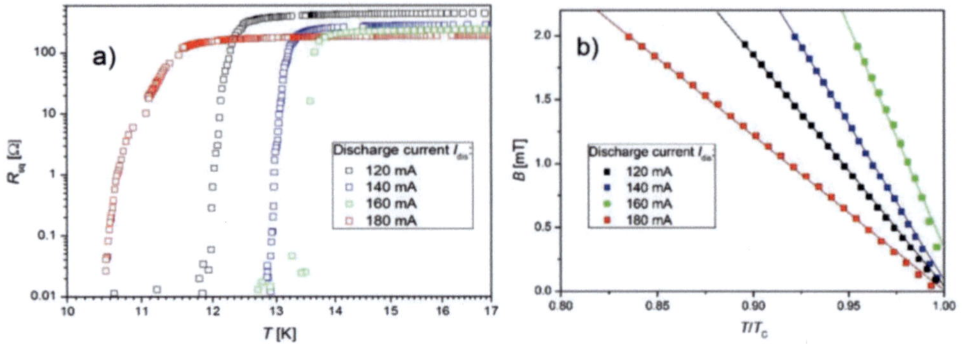

Figure 3.2: a) Dependence of sheet resistance on temperature for films deposited at currents indicated in the legend. b) The temperature dependence of second critical magnetic field B_{C2} for four 5 nm thick NbN films deposited at the sputter currents indicated in the legend. The solid lines are linear fits to the data with the exclusion of points very close to T_C. The derivative of the second critical magnetic field is obtained from the linear part of the dependence.

Since the value of *RRR* and ρ variates for NbN depending on the amount of impurities and other crystallographic defects, it serves as an index of the purity and overall quality of a film.

The temperature dependence of the second critical magnetic field $B_{c2}(T)$ was measured near T_C in an external magnetic field applied perpendicularly to the film surface (shown in Fig. 3.2b). The second critical magnetic field at $T = 0$ K is calculated in the dirty limit to $B_{C2}(0)$ for all films by the following formula from [73]:

$$B_{c2} = 0.69T_c\left(\frac{dB_{c2}}{dT}_{T\to T_c}\right) \quad (3.1)$$

The derivative of the second critical magnetic field dB_{c2}/dT is taken from the linear part of the dependence in vicinity of T_C. Calculated B_{c2} values are presented in Table 3.1.

With increase of discharge current, the square resistance of films decreases linearly. Contrary to R_{sq}, the residual resistance ratio increases continuously with increase of I_{dis}. A similar dependence was found in the calculated specific resistance at 25 K. The critical temperature as well as the second critical magnetic field and its derivative maximize at $I_{dis} = 160$ mA and, then drop to lowest values.

Table 3.1: Parameters of 5 nm thick NbN films deposited with varying chemical compositions.

I_{dis}, mA	R_{sq}, Ω	*RRR*	ρ_{25}, μΩ×cm	T_C, K	ΔT, K	B_{c2}, T	dB/dT, T/K
120	470	0.85	235	11.9	1.31	12.3	1.5
140	301	0.94	151	12.9	1.09	17.3	1.94
160	251	0.98	126	13.55	0.65	22.6	2.42
180	195	1.05	98	10.6	1.41	8.1	1.28

The value of the coherence length at zero temperature is calculated according to:

$$\xi(0) = \sqrt{\frac{\Phi_0}{2\pi B_{c2}(0)}}, \tag{3.2}$$

where Φ_0 is flux quantum.

Calculation of the electron diffusion coefficient *D*, has been done by taking the linear part of the temperature dependence of B_{c2} near T_C and the following expression:

$$D = -\frac{4k_B}{\pi e}\left(\frac{dB_{c2}}{dT}_{T\to T_C}\right)^{-1} \tag{3.3}$$

The magnetic field penetration depth at zero temperature is estimated as:

$$\lambda(0) = \sqrt{\frac{h\rho_{25}}{\pi\mu_0\Delta(0)}}, \tag{3.4}$$

where μ_0 is vacuum permeability constant, $\Delta(0) = 2.05\,k_B T_C$ is the superconducting energy gap found in [74] for NbN film of similar characteristics.

The density of electron states N_0 is described by Einstein relation:

$$N_0 = \frac{1}{e^2 D\rho} \tag{3.5}$$

The calculated values using the material parameters determined above, are given in Table 3.2.

Table 3.2: Extracted characteristics of 5 nm NbN films deposited at different discharge currents.

I_{dis}, mA	ξ (0), nm	D, cm^2/sec	λ (0), nm	Λ, μm	N_0 $C^{-1}cm^{-3}$
120	5.1	0.73	432	74.7	2.3×10^{47}
140	4.4	0.57	332	44.1	4.5×10^{47}
160	3.8	0.45	295	34.8	6.9×10^{47}
180	6.3	0.85	292	34.1	4.7×10^{47}

3.1.2 Influence of chemical composition on optical properties of films

The optical properties of NbN films are measured with a LAMBDA 1050 spectrophotometer. The substrates with the films are illuminated from the front side. An absorption spectrum of the NbN films are shown in Fig. 3.3.

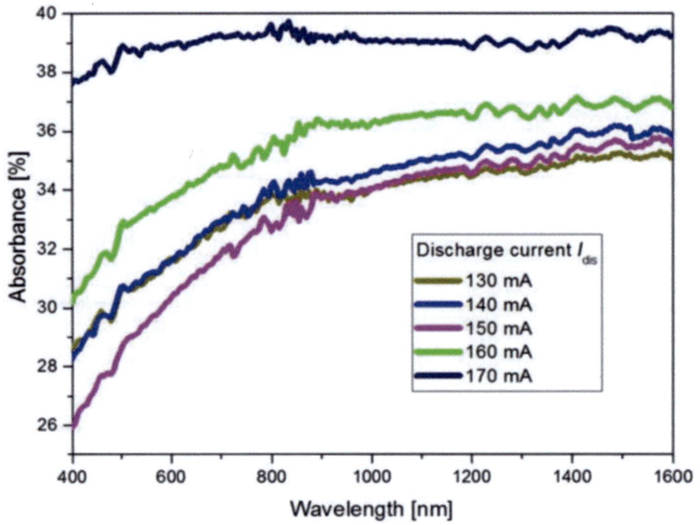

Figure 3.3: Spectral dependence of absorbance of NbN films deposited at different discharge currents Idis indicated in the legend.

The absorbance is obtained as 1- t – r, where t and r are transmittance and reflectance, respectively. For all films except one deposited at 170 mA, the absorbance exhibits an increase in the range of wavelength from 400 to 850 nm and remains almost constant in the infrared range up to 1600 nm. The absorbance of film deposited at 170 mA is almost constant in the whole spectral range of measurement. Nevertheless, the absolute value of absorbance is about 39%, while absorption of other films increases up to 34-36%.

Fig. 3.4 represents results of measurements of reflection and transmission at 1500 nm wavelength. Both characteristics are dropping with an increase of discharge current. The growth of residual-resistance ratio with increase of I_{dis} indicates a transition of the temperature dependence from semiconductor-like to metallic-like behaviour. At the same time, the density of electron states grows in films deposited in the Nb phase. The growing number of electrons increases the probability of absorption and reflection of photons and thus, the corresponding transmission coefficient decreases. The change of reflection of films can be associated with an increase of grain size. It was shown [75] that grain size correlates with specific resistivity of films. The higher variation of ρ in temperature range of deposition results in smaller size of the grain. Surface and grain boundary electron scattering contribute significantly to the resistivity as the dimensions of polycrystalline metallic conductors are reduced to, and below, the electron mean free path, $l \ll \xi$. Grain boundary scattering is found to provide the strongest contribution to the resistivity increase.

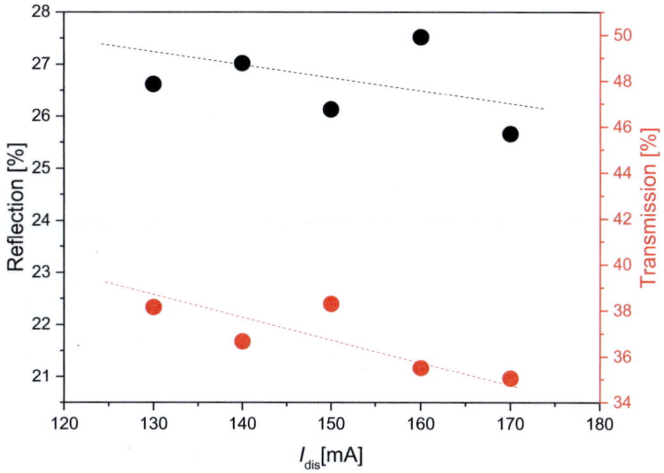

Figure 3.4: The measured reflection (black points) and transmission (red points) of 5 nm NbN films at 1550 nm wavelength. The points are plotted for films deposited at different discharge currents.

As mentioned in the beginning of this section, the *DE* of a single-layer detector is limited by the absorbance of film. In the standard optical setup for SNSPDs, the device is directly illuminated form the surface of nanowires on the substrate. If photons are collected from the backside of the detector area, the obstacle to the absorption could be overcome due to different optical path length.

In order to prove this assumption, a 5 nm thick NbN film deposited on one-polished side sapphire was optically characterized for both front and backside illuminations. For both configurations, the absorbance exhibits an increase in the range of wavelength from 400 nm to 850 nm and remains almost constant in the infrared range up to 1600 nm (Fig. 3.5a). The absorbance of films remains almost constant for front side and rear side illumination at wavelengths >1000 nm. It begins to decrease around 1000 nm that corresponds to the RMS roughness of the unpolished side of the substrate. The difference in the absorption is found to be 22% higher at visible light and 50% at mid-infrared light in case of backside illumination. The explanation of better absorption of films illuminated at back side is straightforward. For the frontside illumination, the light just passes vertically through the NbN layer, so the optical path length is determined only by NbN layer thickness. For backside illumination, light is first diffused by the roughened backside, so that the light travels with different propagation angles (Fig. 3.5b).

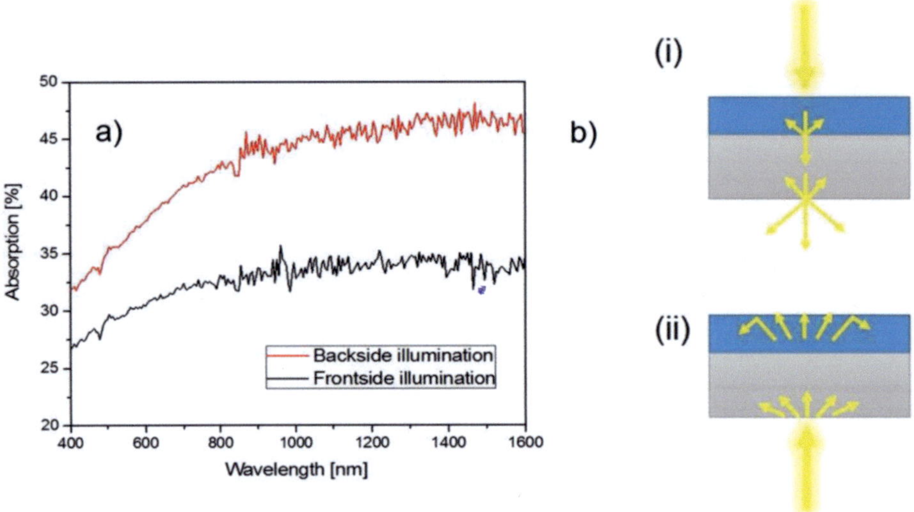

Figure 3.5: a) Absorption spectra of a 5 nm thick NbN film on one-side polished sapphire substrate illuminated from the rear side (red curve) and from the front side (black curve). b) The sketch of light distribution in the NbN-sapphire (blue-grey) system for frontside (i) and backside (ii) illumination.

The consequence is that the average optical path length inside the NbN layer is enhanced compared to the frontside illumination case (optical path length > NbN layer thickness), so the probability of absorption is higher. Furthermore, the total internal reflection of the NbN film reduces the absorbance of the film at frontside illumination due to a limited value of propagation angle in the film. When light travels from a medium with a higher refractive index (NbN film) to one with a lower refractive index (sapphire), then light can be reflected at some critical angle described in frame of Snell's law. If the light is scattered and then propagates inside the NbN film with an angle higher than 11.5 degrees for given film at mid-infrared light, it will be completely reflected back to NbN film. The refractive index of sapphire (1.8-1.73 for wavelengths 400-1600 nm) is lower than NbN (2-5 for wavelengths 400-1600 nm), the backside reflection decreases compared to the front side illumination.

Furthermore, it was assumed that the morphology of the substrate surface influences on absorbance as well. With the same measurement, absorbance of two 5 nm NbN films on substrates with polished and non-polished back side is estimated in a wavelength range between 400 nm and 1600 nm. The difference of the absorption between the films is about 10% in the whole spectral range (Fig. 3.6). The scattering is dependent on the backside roughness of the substrate.

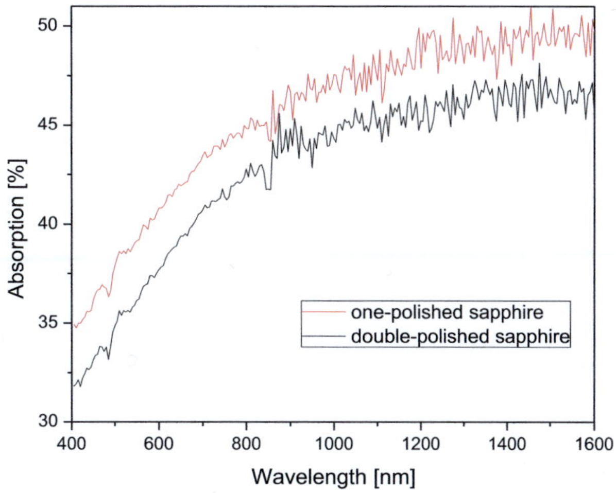

Figure 3.6: Absorption spectra of 5 nm thick NbN film on free standing one- (black curve) and double-side (red curve) polished sapphire substrate illuminated from the rear side.

In conclusion, the absorbance of NbN film grows with the variation of material composition towards Nb phase. The material becomes more optically dense as the niobium content is increased, leading to more favorable absorption conditions. The absorbance can be further increased by changing the optical coupling to backside illumination and using one-side polished sapphire as a substrate for deposition of NbN film. The absolute value of absorbance was increased up to 50% in comparison to the value of NbN film mounted in the standard optical coupling.

3.1.3 Thin NbN films with AlN protection layer

The proximity effect (described in section 2.2.2) in the NbN films is a limitation of the maximum of superconducting properties. It's theoretically explained in the frame of normal conductor-superconductor-normal conductor (NSN) layer systems, but it still didn't demonstrate experimentally. The normal conductors are results of two following processes: 1) the deposition of the first few atomic layers of the film which are most strongly influenced by the lattice mismatch of the substrate and the film; 2) oxidation when the films are exposed to air.

In the following section a NbN/AlN bi-layer system is considered as possible method for preventing formation of upper normal conducting layer. In order to avoid oxidation of the surface of NbN films, the NbN/AlN bi-layer should be sputtered in-situ. The conditions of deposition of AlN have to be optimized. Contrary to NbN film deposition, the discharge characteristic is qualitatively different for the sputtering process for AlN (Fig.3.8). When nitrogen gas is introduced into the chamber, voltage on magnetron drops significantly due to larger secondary electron emission coefficient for AlN than for Al. The nitride film begins to form on the target with corresponding reduction of sputtering of Al. The transition from a metallic to a fully nitrided target surface occurs in a short time. The optimum point at 400 mA (indicated by arrow on Fig.3.8) for deposition of AlN was found early. However, the optimal temperature range for deposition of AlN remains uncertain.

To identify the proper temperature for deposition of a protection layer of AlN, several NbN/AlN bi-layers are deposited with variation of heating: room temperature, 170°C, 775°C. A thin 4 nm NbN film is deposited on one-side polished, R-plane cut sapphire substrates by reactive magnetron sputtering of a pure Nb target in an argon and nitrogen gas atmosphere. Partial pressures of argon and nitrogen were $P_{Ar} = 1.4 \times 10^{-3}$ mbar and $P_{N2} = 0.4 \times 10^{-3}$ mbar, respectively. During deposition, the substrate was placed without being thermally anchored on the surface of a copper holder, which in turn was placed on a heater plate. The plate was kept at a temperature of 775°C during the deposition. The deposition rate of NbN was 0.037 nm/s at a discharge current of 120 mA.

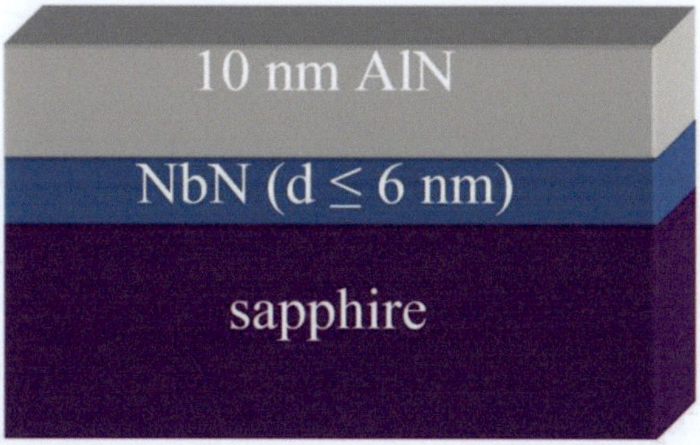

Figure 3.7: NbN/AlN bi-layer system for improving superconducting and electrical properties of deposited NbN film.

Afterwards, the 10 nm thick AlN is deposited *in-situ* at the following pressures: $P_{Ar} = 3 \times 10^{-3}$ mbar and $P_{N2} = 0.4 \times 10^{-3}$ mbar. The deposition rate is about 0.08 nm/s for given conditions.

The measured $R(T)$ dependences of all deposited bi-layers is compared with the temperature dependence of resistance of NbN without a AlN layer. The obtained results are presented in Fig. 3.9.

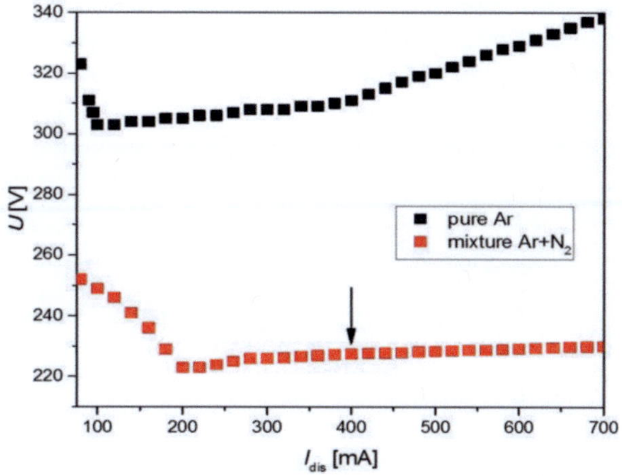

Figure 3.8: The discharge characteristic of reactive magnetron sputtering of AlN.

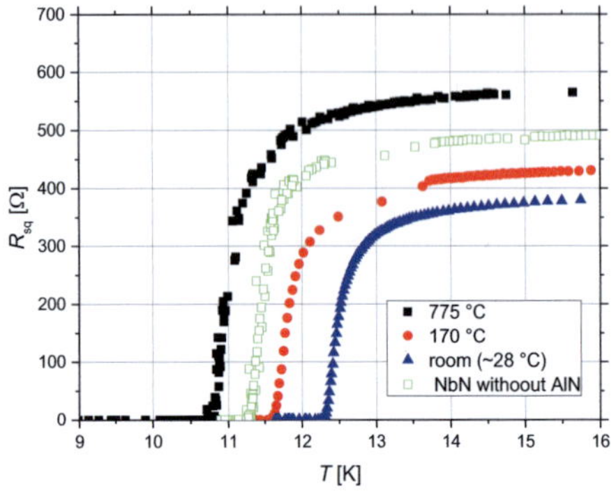

Figure 3.9: *R*(*T*)-dependence of NbN films with AlN protection layer deposited at different temperatures indicated in legend.

Thin 4 nm NbN film without protection layer has a critical temperature 11.3 K (green curve). The deposition of 10 nm AlN layer at room temperature provides higher T_C up to 12.3 K (blue curve). Although heating of the holder to 170°C leads to higher T_C (11.6 K) (red curve) than without AlN, further increase of heating temperature results in film degradation (black curve).

Beside the critical temperature, square resistance and width of transition vary with change of deposition temperature. The experimental points are plotted in Fig. 3.10. The narrowest transition is found to be 1.05 K for the film with room temperature deposited AlN layer. With increase of heating temperature, the width of transition becomes larger. It provides inhomogeneous areas within the film which can be associated with interdiffusion processes of AlN and NbN films at high temperatures. A growing square resistance of films with same nominal thickness in temperature range confirms this assumption.

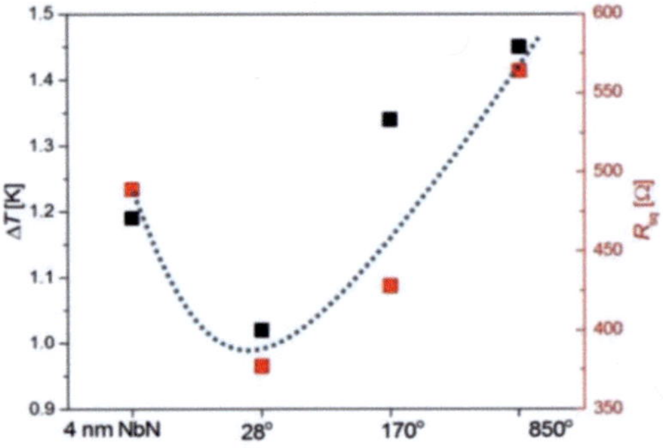

Figure 3.10: The dependence of width of transition (black points, left axis) and square resistance (red points, right axis) on different temperatures of deposition of AlN and without AlN layer.

To investigate the influence of the protection AlN layer deposited at room temperature on superconducting properties of NbN, films are deposited with variation of thickness from 2 up 6 nm with and without AlN. Sputtering of AlN is preferable without heating. As expected, the width of superconducting transition increases with reduction of thickness (Fig.3.11). The assumption that T_C of NbN can be increased by *in-situ* deposition of AlN layer is experimentally confirmed on films with different thicknesses. The difference in T_C between NbN single- and NbN/AlN double-layers is associated with the suppression of oxidation process of the protected surface of NbN film. The increased (about 1 K higher) critical temperature of NbN film broadens the opportunity for making detectors with small dimension.

In conclusion, the electrical and superconducting properties of films with different stoichiometry for fabrication of single-photon detectors was evaluated. The characteristics of films were compared and perspectives for single-photon detector.

The superconducting and material properties of thin NbN films limit the performance of fabricated single-photon detectors. The critical temperature of the final device should be at least two times higher than a working temperature. The film deposited at 180 mA has rather low T_C = 10.6 K. Taking into account suppression of T_C due to the technological process, the critical temperature of the final detector will be further reduced down to operation temperature at 4.2 K. Contrariwise, the minimal photon energy, which is required to generate a hot-spot of sufficient size, depends on superconducting energy gap Δ. The optimum of deposition lies in between two limitations.

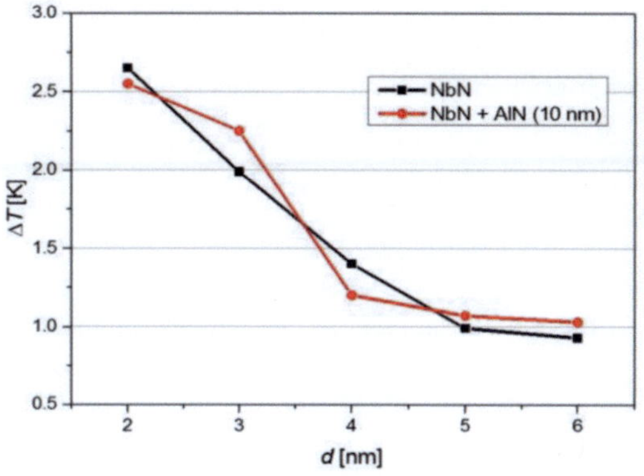

Figure 3.11: Width of transition versus thickness for films with (red) and without AlN layer (black) on top.

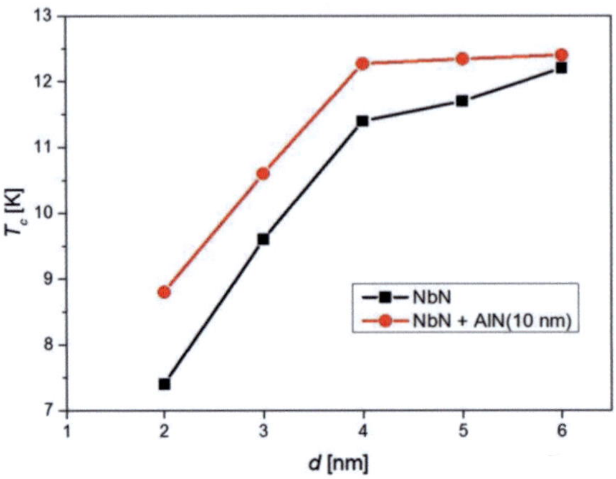

Figure 3.12: Critical temperature of films with variation of thickness for NbN single layer (black points) and NbN/AlN bi-layer (red points).

From another point of view, the hot-spot diffusion theory (described in section 2.2.1) predicts broadening of the hot-spot response spectrum in strips with low density of electronic states, N_0. The density of electron state, which depends on out-diffusion of quasiparticles and resistive properties of films, can be adjusted by variation of stoichiometry of the sputtered film. However, when the diffusion process is too fast, the collapse of the hot spot occurs on a shorter time scale than the redistribution of the applied current and the formation of a normal belt across the nanowire. It was also demonstrated that detectors which are made from films deposited in Nb phase with a lower value of ρ has relatively higher ratio of the critical current to the calculated limit. Thus, films deposited at a current of 120 mA with high specific resistivity and diffusion of quasi-particles, are not appropriate for the fabrication of high-efficient infrared detectors.

Films with similar characteristics were deposited at I_{dis} 140 and 160 mA. The slightly higher T_C of films deposited at 160 mA allows making nanowires with a smaller cross-section.

3.2 Nanowire patterning

Superconductivity and geometrical effects in low-dimensional structures close to mesoscopic and quantum limits, i.e. in structures where the thickness, width and length separately or together become comparable and smaller than the coherence length ξ and magnetic field penetration depth λ, are in focus of numerous theoretical and experimental works. This interest is stimulated by the fundamental importance of the problem itself as well as by the variety of applications of thin-film superconducting nanostructures.

In the following section, a method to improve the quality of nanowires is presented in details. The experimental results obtained by specific lithography processes and theoretical aspects will be analyzed and discussed thoroughly.

3.2.1 Positive- and negative-PMMA lithography

PMMA resist is a well-known positive-tone resist which is attractive for users due to easy handling, high stability, reproducibility of lithographed structures, and high resolution. PMMA resist is available with different sensitivities. The required thickness of the resist layer can be easily achieved by varying the speed of spinning and/or the amount of solid content in the resist. PMMA itself and the required developer and stopper are water-free materials that are preferable for pattering films from water-sensitive compounds. A disadvantage of PMMA resist is the relatively high temperature (between 150 and 190°C), which is required to bake the resist after spinning. Such high temperatures stimulate the diffusion of oxygen that increases its penetration depth into the film where, for films from Nb compounds, oxygen deteriorates or suppresses superconductivity. Furthermore, this resist is moderately stable against plasma assisted etching processes that limits its applicability especially in the case of thin resist layers, which are required for writing ultimately small features. Although PMMA electron-beam resist was originally introduced as a positive tone resist, it also can be used for negative lithographic processes.

When a primary beam of electrons with energies ≈10 keV (far below the threshold for the displacement of carbon atom [76]) enters the PMMA resist and a substrate, it produces low-energy secondary electrons (SE), which are mainly responsible for the scission process of the PMMA polymer chain [77]. A reduced molecular weight of the PMMA resist, which is exposed with a dose in the range of 100 $\mu C/cm^2$, makes it soluble in solvents with a high enough activation energy [78]. In this case PMMA acts as a positive-tone resist.

At high exposure doses (≥ 1 mC/cm^2) PMMA chains decompose into very short low-molecular-weight fragments, which start to form a dense carbonized film. The structures made of this film are insoluble in a standard PMMA developer and even in acetone due to a cross-linking and formation of the covalent bonds between fragments. In this case PMMA acts as a high-resolution negative-tone electron-beam resist [79].

Two identical NbN films were patterned simultaneously one by the positive and another by the negative process in order to eliminate different aging degrees of the films. The substrates were spin coated with PMMA 950k resist to a layer thickness of 95 nm. In order to minimize degradation of the films, the resist was baked at the lowest recommended temperature of 150°C for 5 minutes.

The layout was the same for all samples in both the positive- and negative-PMMA series and represented a straight nanowire, which was embedded between small contact pads. In order to avoid current crowding [60] at the steps from the pads with a width of a few tens of micrometers to the nanowire, they were rounded off with a radius r ≈ 4 μm. In each series, the width of nanowires was varied between 50 nm and 20 μm. The actual widths of nanowires and strips were measured using scanning electron microscopy (SEM).

In case of "standard" positive-PMMA process, two separate islands were exposed by 10 kV electron beam with the dose about 100 μC/cm². The islands were separated by a slit which in its middle had the width equal to the design width of the nanowire and grew at edges to encompass rounded steps to contact pads. After development in the standard developer for 30 sec (30% MIBK in 2-propanol at 23 °C) and rinsing in the 2-propanol stopper, the exposed areas were removed. Unexposed resist between islands remained on the surface of film (Fig. 3.13a) and protected the film during the subsequent ion-milling process.

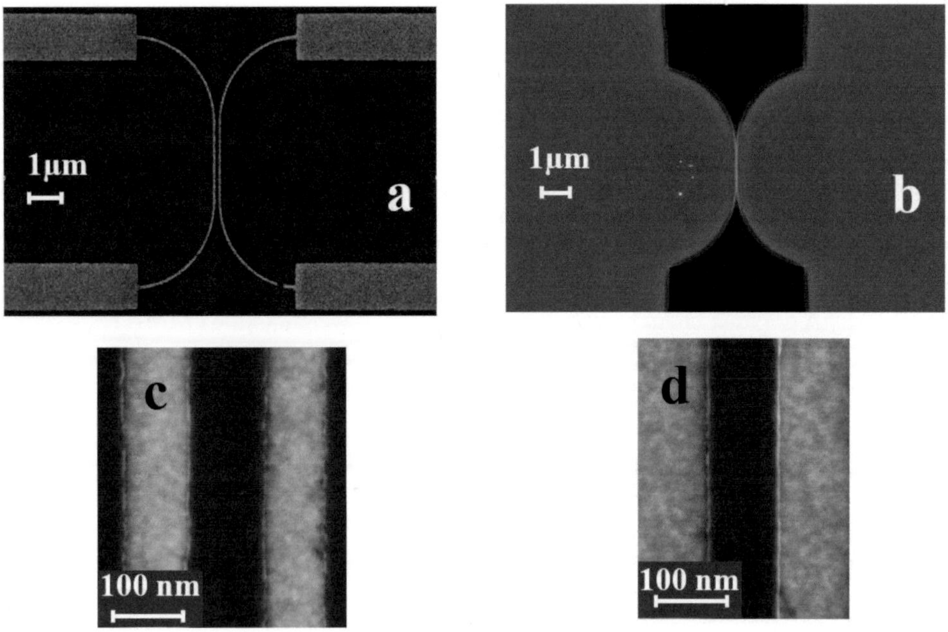

Figure 3.13: SEM images of nanowires which were prepared by the positive-PMMA (a, c) and the negative-PMMA (b, d) lithography: a) and b) – after development of the resist (dark areas – PMMA resist); c) and d) – central parts of nanowires prepared by two different lithography after ion milling and stripping off the resist (dark areas – NbN film).

Large contact pads sized to a few millimeters for ultrasonic bonding were prepared by photolithography with a mask which additionally protected the already patterned nanowire and small pads during the second etching step and separated samples from each other making them ready to measure.

In case of "standard" positive-PMMA process, two separate islands were exposed by 10 kV electron beam with the dose about 100 μC/cm². The islands were separated by a slit which in its middle had the width equal to the design width of the nanowire and grew at edges to encompass rounded steps to contact pads. After development in the standard developer for 30 sec (30% MIBK in 2-propanol at 23 °C) and rinsing in the 2-propanol stopper, the exposed areas were removed. Unexposed resist between islands remained on the surface of film (Fig. 3.13a) and protected the film during the subsequent ion-milling process. Large contact pads sized to a few millimeters for ultrasonic bonding were prepared by photolithography with a mask which additionally protected the already patterned nanowire and small pads during the second etching step and separated samples from each other making them ready to measure.

In the negative-PMMA process, the exposure dose of the resist was increased by two orders of magnitude to reach 10 mC/cm² while the energy of electrons was kept unchanged. In contrast to the positive-PMMA process described above, here the electron beam exposed only nanowire and small pads (Fig. 3.13b) and after development these areas remained on the film surface. The pattern was developed in acetone for 1.5 min and then rinsed in 2-propanol. Right after that, large contact pads were prepared by standard photolithography with a mask which left area with the nanowire and rounded steps under negative-PMMA open but overlapped with the small pads.

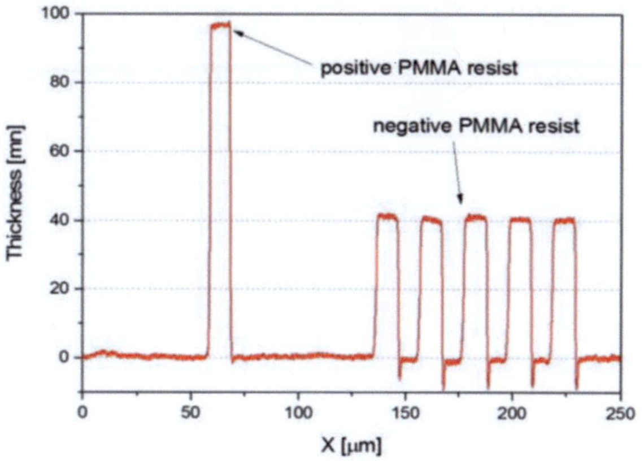

Figure 3.14: Probing of PMMA resist after exposure of two techniques: positive-PMMA lithography and negative-PMMA lithography. Measured peaks are indicated by arrows.

After development of the photoresist, the complete image, which contains the central part with negative PMMA and large contact pads, was transferred into the NbN film by ion-milling process.

It has to be noted that the thickness of PMMA resist at the areas, which were exposed with the large dose of 10 mC/cm^2, shrunk from 95 nm to about 55 nm (this value is measured right after exposure) and remained unchanged after the development in acetone (Fig. 3.14). No measurable changes were observed in the thickness of PMMA resist exposed by the low dose of 100 μC/cm^2. The etching rate of the negative-PMMA resist was found to be about 2.7 nm/min. This rate is comparable to the etching rate of NbN by Ar ions with an energy 200 eV and current density ≈ 1 mA/cm^2 at 10° incident angle. At the same etching conditions, the etching rate of positive-PMMA resist was almost 2.5 times higher (≈ 6.7 nm/min) (Fig. 3.15).

After etching the residual resist (both positive and negative) was removed from the surface of NbN film using a combination of warm acetone, ultrasonic shaking and a gentle mechanical brushing.

In turn, the increase in writing time requires additional efforts to get long-term stability of electron-beam parameters and long-term suppression of external acoustic, mechanic, and electro-magnetic interferences disturbing the lithography apparatus.

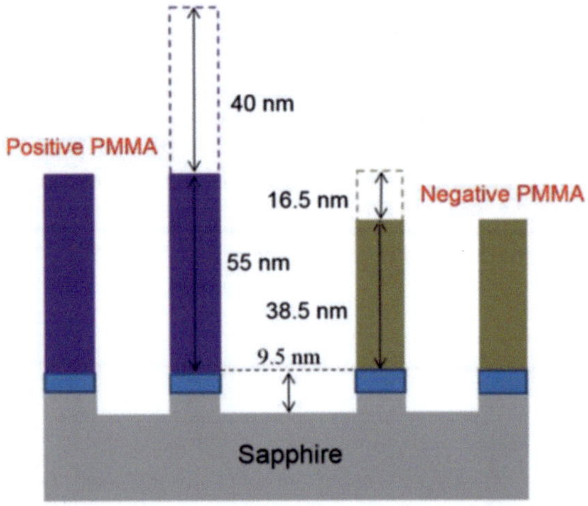

Figure 3.15: Schematically illustrated profile of ion-etched NbN film with positive- and negative-tone resists on it after etching process. The total etching time is 6 min.

To estimate the quality of edges made by positive- and negative-PMMA lithography, a statistical analysis is performed of the edge roughness of nanowires. In Fig. 3.13 there are two SEM images of nominally 100-nm wide nanowires after ion milling. Nanowires were prepared by positive (Fig. 3.13c) and negative (Fig. 3.13d) electron-beam lithography of PMMA resist. The rest of resist was stripped-off and samples were thoroughly cleaned. For the analysis of the nanowire geometry, a high-resolution field-emission SEM equipped with the highly-efficient in-lens detector of secondary electrons (SE) is used. This detector is sensitive mainly to secondary electrons, which are generated in vicinity of the primary beam at the very surface of a sample ("SE type-I"). Therefore, the in-lens detector ensures spatial resolution of about 1-2 nm. The raster scan was aligned along the length of nanowires to minimize low-frequency- noise contribution to the image. A set of images with different magnification were acquired with the nominal resolutions better than 0.5 nm/pixel.

The data from the SEM images is used to extract the mean value and the standard deviation of the width of nanowires fabricated by the negative- and positive-PMMA lithography. In order to do that, the recorded raster scan which is 2D-array with 255 gradations of the SE-signal intensity per pixel is used. The lines, which correspond to the edges of the nanowire, were defined using the Canny edge-detection algorithm [80]. The local width of the nanowire is defined as a distance between detected edges in the direction perpendicular to the line of the raster scan. The standard deviation was obtained from a Gaussian fit of the histogram of the local width-wire. The standard deviation for positive-PMMA nanowires is $\approx$ 5 nm, while for negative-PMMA nanowires it is smaller $\approx$ 2-3 nm. An example of the statistical analysis is presented on Fig. 3.16. There are several factors which may result in increased roughness of the width of the positive-PMMA nanowires. The presence of the solvent residues in the positive PMMA can lead to a non-uniform development process. The etching rate of PMMA by accelerated Ar ions could be also different for areas with trapped solvent residues. During ion milling, irradiation by low energy Ar ions could lead to carbonization of the resist layer in that structures from the positive PMMA will contract non-uniformly. The latter might also lead to the edge damage of the NbN nanowire.

In conclusion, negative-PMMA lithography offers the following advantages over positive-PMMA lithography. The negative-PMMA resist is more stable which is seen in more than two times smaller etching rate. The 50 nm thick layer of the negative-PMMA resist protects the film during etching process as good as almost the twice thicker layer of the positive-PMMA resist. In electron-beam lithography, smaller thickness of the resist layer, in general, allows for writing smaller pixels in a reproducible manner.

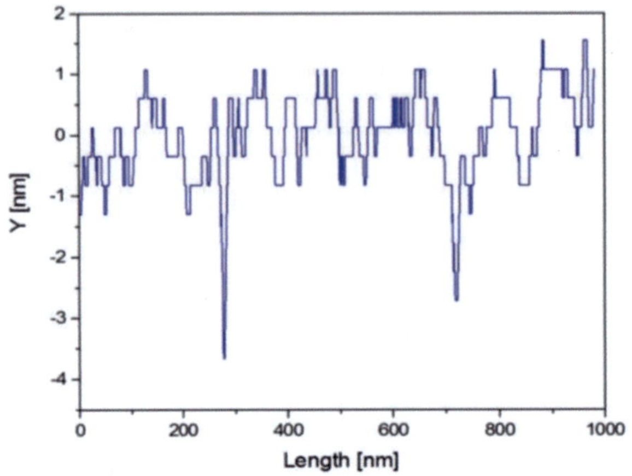

Figure 3.16: Example of statistical analysis of the edge roughness of nanowires.

3.3 Patterning of SNSPDs

The common patterning process of SNSPDs from a NbN single-layer film includes only two steps. The single-spiral technology implies fabrication of multilayer devices with additional steps of technology. Because the single-spiral technology is not developed yet, the technological route of multilayer single-spiral SNSPD is required.

In this section, the technology of SNSPD is described for different layouts. The development of SNSPD technology in form of single spiral is presented in detail. Optimization of technological processes for single spiral is explained as well.

3.3.1 Single-layer process

The meander shape is a widely used layout for SNSPD. Recently, the double-spiral geometry was proposed as a detector layout for SNSPD [36]. The double spiral consists of two nested spirals which are connected at the blind spot. In this case, the current flows from the outer contact to the center of the circle and then back to another outer contact. The sharper turn in the center of the double spiral is needed to connect the two spiral arms (Fig. 3.17). Specimens with meander and double-spiral layouts are prepared in two steps. Firstly, the film is patterned by means of e-beam lithography and ion-beam etching to form nanowires. The width and a quality of nanowires are controlled by SEM inspection. Secondly, a coplanar line is made around the wired area by photolithography and reactive-ion etching (Fig. 3.17).

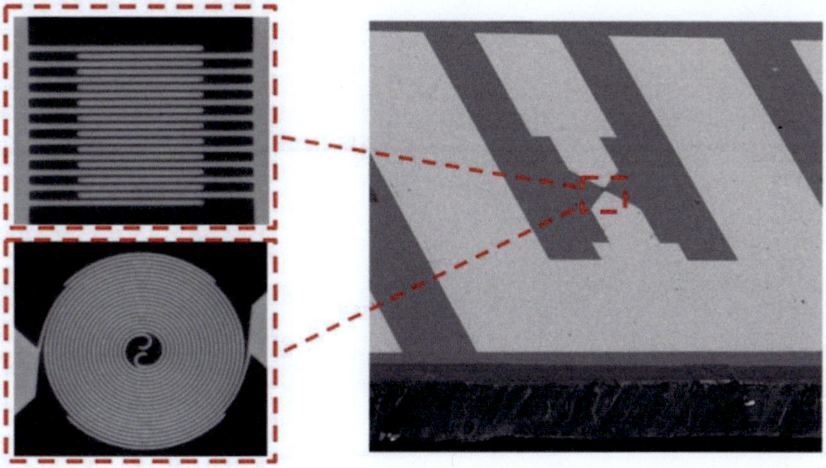

Figure 3.17: Coplanar layout with embedded detector either in form of meander or the double-spiral.

3.3.2 Multilayer single-spiral structures

Preparation of specimens with a single-spiral layout is more complicated. Aside from pattering the film into a spiral, it includes fabrication of an isolating layer between the spiral and the top contact and patterning the top contact itself.

The first step of fabrication is patterning of spirals from thin NbN films, using electron-beam lithography and ion-beam etching techniques. The polymethyl methacrylate (PMMA) resist is used with a thickness of about 65 nm. The NbN film was etched by argon-ion milling ($P_{Ar} = 1.1 \times 10^{-4}$ mbar).

At the second step of patterning, the spiral needs to be isolated from the top except the central pad for leading the current through the spiral. Aluminum nitride (AlN) was chosen as a material for isolation. A circle with diameter larger than the outer diameter of the spiral is exposed in PMMA resist. The 50 nm-thick aluminum nitride film is deposited by reactive magnetron sputtering of pure Al target in a mixture of argon and nitrogen atmosphere at a partial pressure of argon $P_{Ar} = 3 \times 10^{-3}$ mbar and of nitrogen $P_{N2} = 4.5 \times 10^{-3}$ mbar at room temperature. After deposition of aluminum nitride, the sample was put into warm acetone for lift-off. The 50 nm-thick layer of AlN reliably isolates the spiral from contacting with the electrode which is deposited on top at the third step of fabrication. The technological route of single-spiral SNSPD is presented on Fig 3.18.

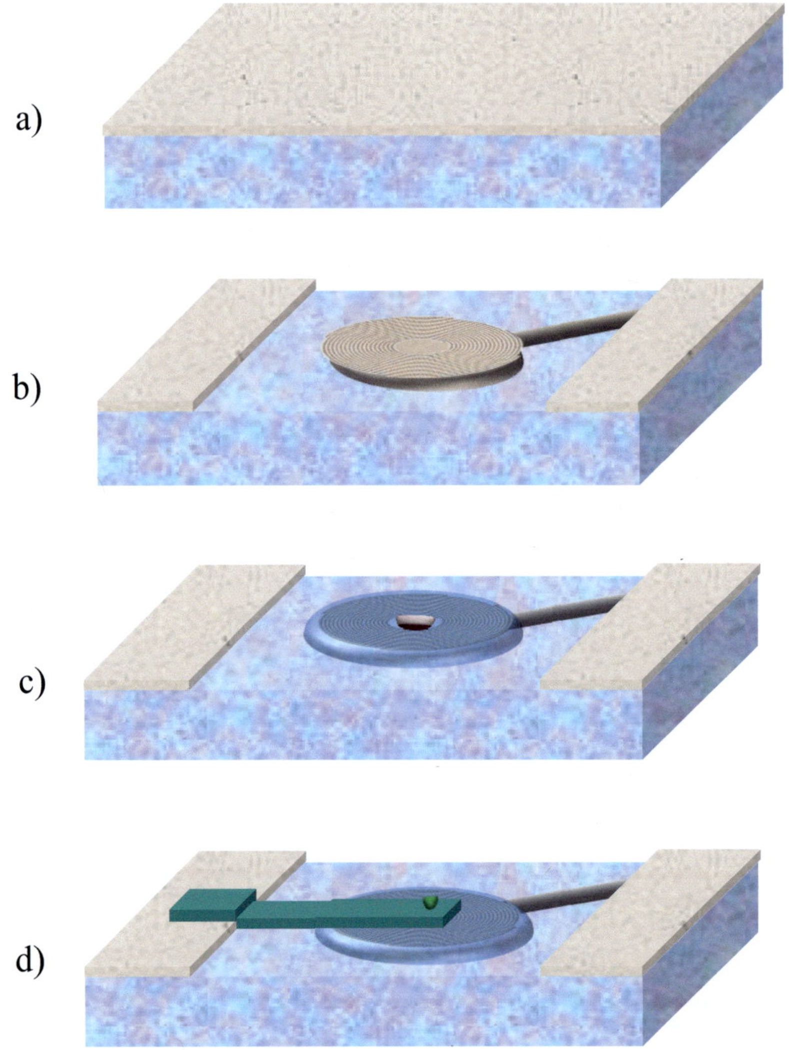

Figure 3.18: Technological route of fabrication of single-spiral detectors: a) Deposition of NbN films on heated sapphire substrate; b) Patterning of single spiral using standard positive e-beam lithography and ion-milling etching; c) Isolating layer of 50 nm thick AlN is made by lift-off; d) Deposition of thick Nb layer as a top electrical contact to the spiral.

Patterning of the top contact pad is the last step of technological route of spiral SNSPDs. The top electrode has to be at least two times thicker than the isolator for proper electrical contact to the spiral. The top electrode was fabricated from a 100 nm thick superconducting Nb film by electron-beam lithography and magnetron sputtering of pure Nb at pressure $P_{Ar} = 5 \times 10^{-3}$ mbar.

At this step, PMMA resist with thickness about 120 nm was used for fabrication. Thicknesses of deposited layers are presented on the sketch in Fig. 3.19. The basic scheme is presented on Fig. 3.20. Two designs of spirals were chosen for SNSPDs: circular spiral (Archimedean or single spiral) and square spirals, sometimes also called Egyptian or Greek spiral. Images of the final structure are displayed in Fig. 3.21 for single spiral (a) and square spiral (b).

It was found that spirals with Nb top electrode deposited without pre-cleaning have an offset resistance below the critical temperature due to a thin normal layer on top of spiral made of NbN. A pre-cleaning process is required for a proper contact of top electrode. By pre-cleaning the oxidized thin layer on top of spiral will be removed with rest organic particles remaining after development.

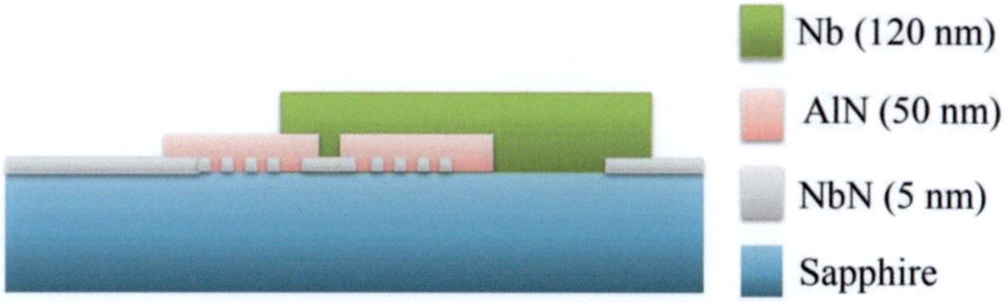

Figure 3.19: Cross-section of single-spiral SNSPD

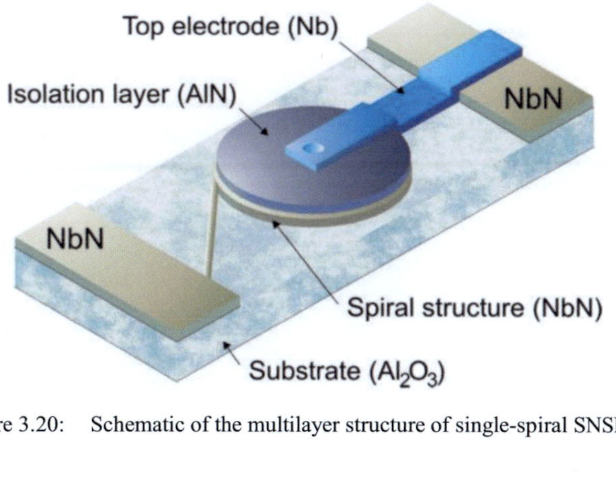

Figure 3.20: Schematic of the multilayer structure of single-spiral SNSPD.

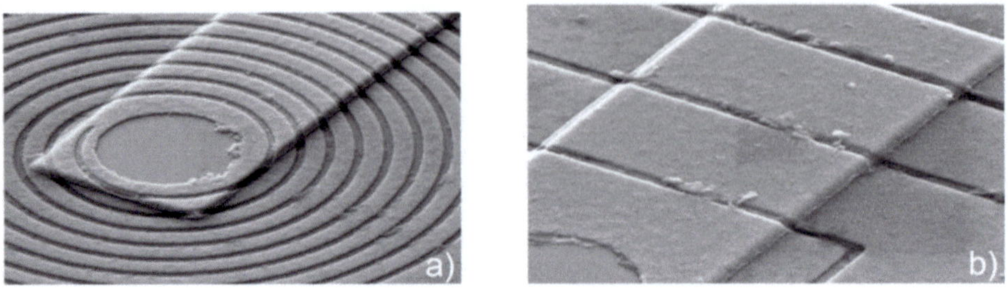

Figure 3.21: SEM images of fabricated detectors of single spiral (a) and square spiral (b).

To investigate the influence of pre-cleaning process on superconducting properties of films, the pre-cleaning time was varied in range from 0 to 20 minutes. During the etching process at the room temperature the current of ion-gun was set to be 10 mA with corresponding pressure of argon $P_{Ar} = 1 \times 10^{-3}$ mbar. The etching is 0.065 nm/min for given conditions.

Results of $R(T)$ measurements of films after pre-cleaning processes are presented on Fig. 3.22. With increase of cleaning time, resistance of films with same nominal electrical properties grows from 1400 up to 2200 Ω. At the same time, the critical temperature stays the same for films with cleaning time not longer than 5 minutes. Further increase of cleaning time results in etching of the NbN film and corresponding reduction of the critical temperature.

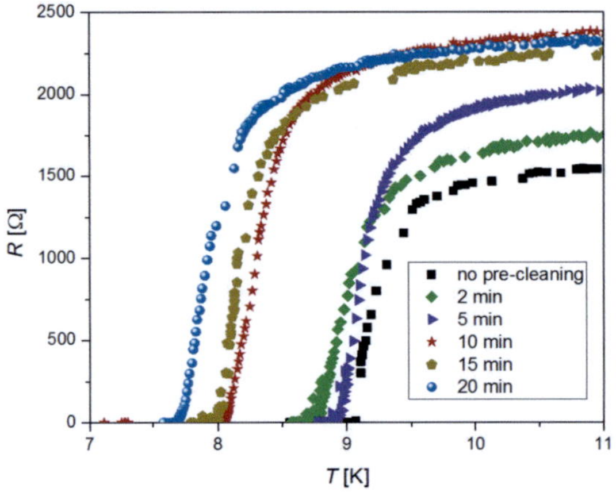

Figure 3.22: $R(T)$-dependence of superconducting 3.5 nm thin NbN films with different time of ion-milling.

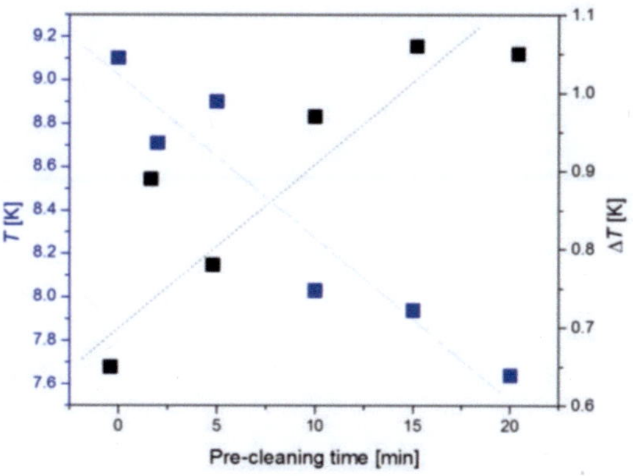

Figure 3.23: The dependence of the critical current (blue dots) and the width of transition (black dots) on pre-cleaning time.

The width of transition increases with decrease of the critical temperature from 0.6 up to 1.05 K (Fig. 3.23). The critical temperature drops from 9.1 down to 7.6 K. Thus, the pre-cleaning process which is required for proper electrical contact between spiral and top electrode should be shorter than 5 minutes.

The spirals are designed using mathematical equations (Fig. 3.24). The Archimedean spiral can be described with polar equation as:

$$r = a\theta \ , \tag{3.6}$$

where r is the radial distance, a is parameter which determines how tightly the spiral is "wrapped" (constant parameter for Archimedean spiral is $2\pi a$), θ is the polar angle.

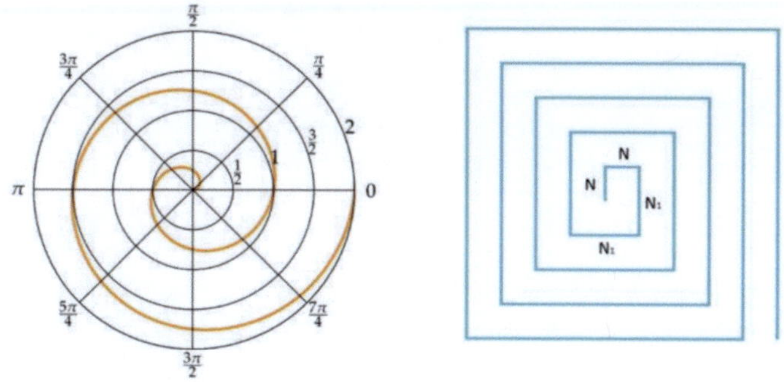

Figure 3.24: General view of one arm of an Archimedean spiral (left image) (represented on a polar graph) and square spiral (left image).

Then, curvature κ of Archimedean spiral is:

$$k(\theta) = \frac{2+\theta^2}{a\left(1+\theta^2\right)^{1.5}} \quad (3.7)$$

The length L of single spiral can be estimated by:

$$L(\theta) = \frac{1}{2}a\left[\theta\sqrt{1+\theta^2}\,\ln\left(\theta+\sqrt{1+\theta^2}\right)\right] \quad (3.8)$$

The square spiral is described by following equation:

$$L = N + N + N_1 + N_1 + \ldots + N_n + N_n \quad (3.9)$$

$$N_1 = N + 2W + a$$
$$N_2 = N_1 + 2W + a$$
$$\ldots$$
$$N_n = N_{n-1} + 2W + a,$$

where L is a total length of square spiral, N – length of the first segment, W is a width on the line, a – is a gap between two lines. The N defines the size of inner central contact pad.

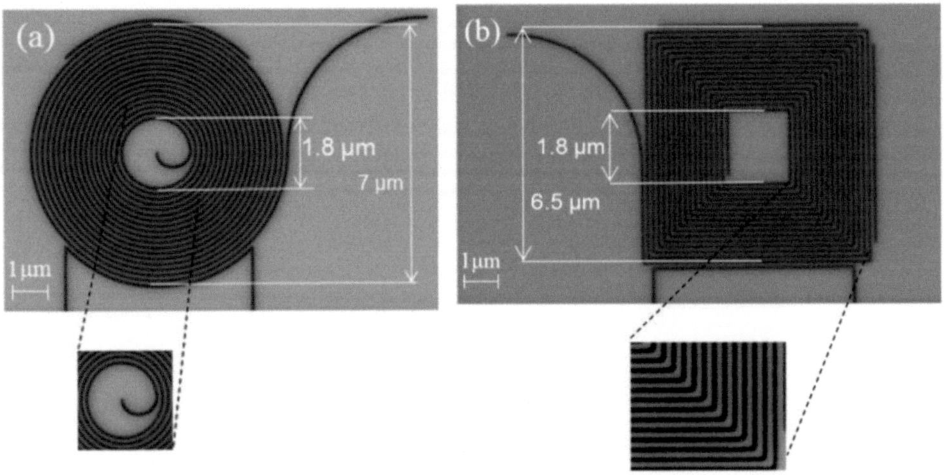

Figure 3.25: SEM images of (a) a circular spiral and (b) a square spiral. Dark color represents strips and surrounding fields from NbN film.

The SEM images of spirals are shown in Fig. 3.25 a, b. Spirals have an inner contact pad in form of a circle (circular spirals) and square (square spirals) in the inner center with diameter 1.8 μm and length of side 1.8 μm, respectively. The outer diameter of the Archimedean spiral is 7 μm. The size of the square spiral is 6.5 × 6.5 μm². The 100 nm width of nanowires, *W*, and 50 nm gap between strips determine the value of the filling factor of the spiral about 70%. A smooth turn is present in the center of single spirals. The geometry of finally fabricated spirals is measured using a scanning-electron microscope (SEM). For single spiral the current flows between the contact outside of the circle and the contact at the blind spot.

The width of nanowires of meander and double-spiral was 100 nm with a 100 nm gap in between, thus the geometric filling factor was 50%. The inner radius of bends in meander is measured to be 35 nm. A meander line fills up a squared area 4 × 4 μm² (Fig. 3.26a). The spiral lines are both confined within the circle with a diameter of 7 μm². The area covered by the spiral has a blind spot in the geometric center of the circle with a diameter either 1.5 μm for the double spiral (Fig. 3.26b). The reference structure was additionally fabricated in form of straight line for DC characterization of the films and, specifically, to obtain the reference value of the critical current density. The dimensions of both geometries are presented on Fig. 3.26.

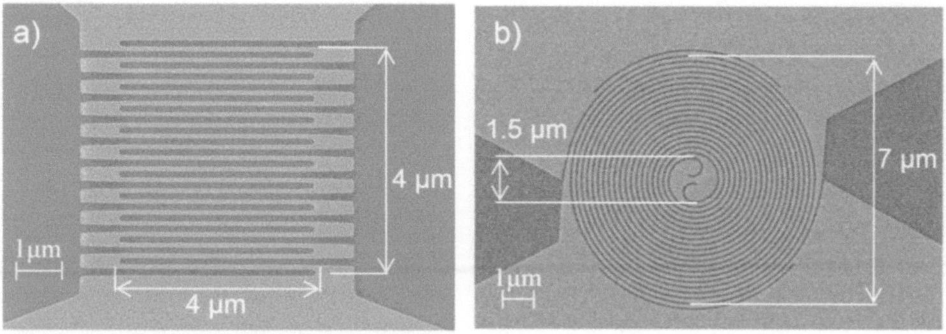

Figure 3.26: SEM images of specimens with different layouts: double-spiral (a), meander (b). The width of nanowires is the same for both layouts ($W = 100$ nm). Dark areas correspond to the surface of the substrate (insulator); bright areas represent NbN lines.

3.4 Summary

This chapter presented a detailed description of methods for improvement of optical and superconducting properties of film and nanostructures. The absorbance can be enhanced by variation of stoichiometry, optical coupling with backside illumination of NbN on one-polished sapphire. Taking all suggestions into account, the absorbance of NbN film can reach 50% in absolute value without using any additional cavity or mirror. The proximity effect is weaker in bi-layer NbN/AlN in comparison to single-layer NbN. It was found that *in-situ* deposited NbN/AlN layers have up to 1 K higher the critical temperature than single-layer NbN with the same nominal thickness below 6 nm. It allows to reduce the cross-section of nanowires for improvement of sensitivity to low-energy photons. The quality of nanowires is improved by using negative-PMMA lithography on standard positive resist. The edge roughness is smaller in nanowires made by negative-PMMA lithography. The higher resolution of the negative-PMMA process allowed to obtain, in a reproducible manner, nanowires with smaller width and smaller edge roughness. The negative-PMMA lithography has a potential in fabrication of nanowires smaller than 100 nm.

The single-spiral technology was developed for two types of detectors: Archimedean spiral and square spiral. All technological steps are optimized to minimum influence the process of making of detectors on superconducting properties of the film.

4 Lateral proximity system (N-S-N) in superconducting nanowire

The current carrying ability of thin films in superconducting state is important for experimental evaluation of their fundamental properties as well as from the application point of view. Therefore, the issue of restricted current has been attracting attention for decades. Reduction of the measurable critical current I_C in superconducting structures with defects belongs to one of fundamental problems of superconductivity (described on section 2.2.2). The experimentally achievable critical current I_C is usually smaller or even much smaller than the depairing current. The critical current is affected by different types of defects such as cross-section variations due to the non-uniformity of the thickness or width of the film, edge defects of nanowires or internal structural defects weakening the superconducting order. The detailed influence of the edge quality on superconducting properties of strips remains unclear. No one theoretical model quantitatively describes the suppression of superconductivity in strips depending on the width of damaged edges. Intuitively, one expects rough edges to cause larger number of defects and correspondingly a reduction of the experimental critical current. The superconducting current circumvents defects. This phenomenon leads to non-uniform current density which peaks at locations adjacent to the defects. Additionally, superconductivity deteriorates at the edges due to mechanical damaging via etching and subsequent oxidation when exposed to air. All those defects reduce the experimental current and make strip characteristics worse.

In the following chapter, the lateral proximity system (N-S-N) [CSB++17] is considered to describe dependencies of superconducting properties of NbN strips on their nominal width W_{SEM}. The width of normal bands at the strip-edges is varied by using two technological approaches [CSB+17]. The effective width of the superconducting core, $W_{SC} < W_{SEM}$, has been extracted from the magnetic field that corresponds to the transition from the Meissner to the vortex state. This occurs when the first magnetic vortex enters the core. By plotting the effective superconducting width to experimental dependencies of basic characteristics of superconducting structures, the behavior in a wide range of width is described. The local density of the critical current in the superconducting core is compared with the theoretical density of the depairing current.

4.1 Superconducting and transport properties of strips

Pattering techniques for variation of width of normal bands

The process begins with reactive magnetron sputtering of NbN on two identical 10 by 10 mm^2 sapphire substrates (the process is detailed described in the section 3.1). The films on both substrates were patterned in parallel by the electron-beam lithography and Ar-ion milling technique. The electron-beam lithography was made over PMMA 950k resist with a thickness about 95 nm. The resist on one of the substrates was exposed with a dose about 100 $\mu C/cm^2$ and developed using standard developer MIBK diluted with 2-propanol (P-chip (strip)). On the second chip the resist was exposed with a two orders of magnitude higher dose about 10 mC/cm^2 (N-chip (strips)). The high difference in the exposure dose resulted in a different profile and hardness of the resist which in turn led to different roughness of edges of final NbN strips: 2-3 nm on the N-chip and about 5 nm for the P-chip. This allows us to obtain two series of NbN strips with two different levels of damaging of their edges and to perform systematic investigation of the influence of these edges on normal and superconducting state properties of NbN strips. More details on the negative- and positive-PMMA technologies can be found in the section 3.2.1.

On each chip there were 20 samples each containing single straight strips of NbN embedded between two contact pads made of the same NbN film. The width of the strips was varied from 50 nm to 20 µm. The smallest width was limited by the smallest feature size which can be reproducibly realized by technology. The value of the Pearl length, $\Lambda = 2\lambda^2/d = 32.9$ µm, was found larger than width of the widest (20 µm) strip. This experiment from a theoretical point of view should ensure a uniform distribution of the supercurrent across the whole width. The samples were designed with a smooth transition (radius of rounding is 5-15 times larger than width of a strip) from strip to contact pads to avoid the current crowding [60]. The nominal width, W_{SEM}, of strips was measured by the scanning electron microscopy.

Characterization of strips

The temperature dependence of resistance of all strips was measured by a four-probe technique. The bias current I_b produces a certain portion of Joule heating determined by its value and resistance of the strip in the normal conducting domain. The narrow strips have small cross-sections with relatively high resistance. On order to prevent self-heating of strips with evident influence of $R(T)$ dependence, the range of permissible currents has to be established. Fig. 4.1 presents results of $R(T)$ measurements of narrowest strip 50 nm with $I_C = 31.2$ µA at different bias current.

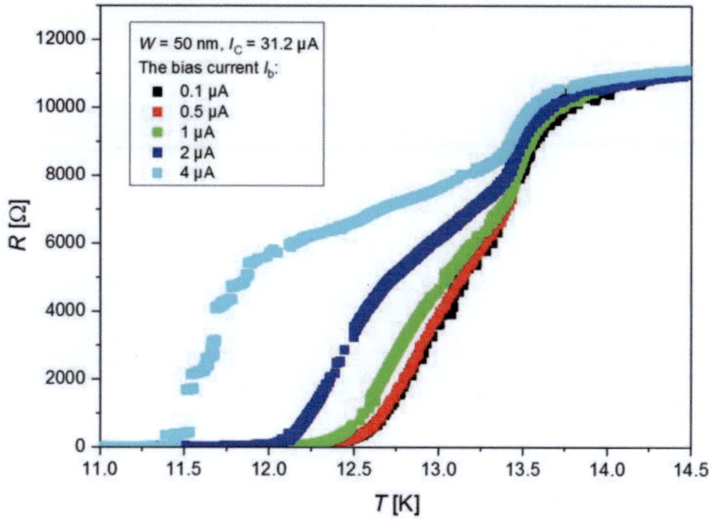

Figure 4.1: Influence of Joule heating on the $R(T)$ shape of a 50 nm strip. The specification of current is indicated in the legend. Joule heating grows with increase of the bias current. The applied current to the strip higher than 1 µA impacts on superconducting transition and temperature.

The transport current is variated in range from 0.1 up to 4 µA. Once set up, the resistance of the strip is measured as a function of temperature. The width of transition stretches at bias currents above 1 µA with reduction of the critical temperature. A further increase of I_b provides strong degradation of measured T_C and overheating of the strip.

A bias current below 1 µA ensures negligible heating and unaffected measured superconducting properties. The lowest bias current value of 0.1 µA is restricted by noise of the measurement system. By establishing of permitted bias currents, the influence of self-heating is excluded from transport measurements of narrow strips.

The temperature dependence of resistance has been measured at temperatures ranging from 4.2 up to 300 K for two series of strips made by positive- and negative-PMMA technologies. Examples of $R(T)$-curves for nano- and microstrips are presented on Fig.4.2. The resistance of strips was varied in the range between 1.7 and 12.3 kΩ by a changing strip width. The transition of the nanostips shows a double step shape which is definitely not related to self-heating. Because strips are embedded in the pair of contact pads with dimensions much larger than width of strips, the sample should be considered as electrical series circuit of resistors. The anticipated transition of contact pads to the superconducting state is observed at non-zero resistance. For example, the reduction of the resistance of a 54 nm narrow strip (red curve on Fig. 4.2) from 12.3 down to 5.8 kΩ was measured.

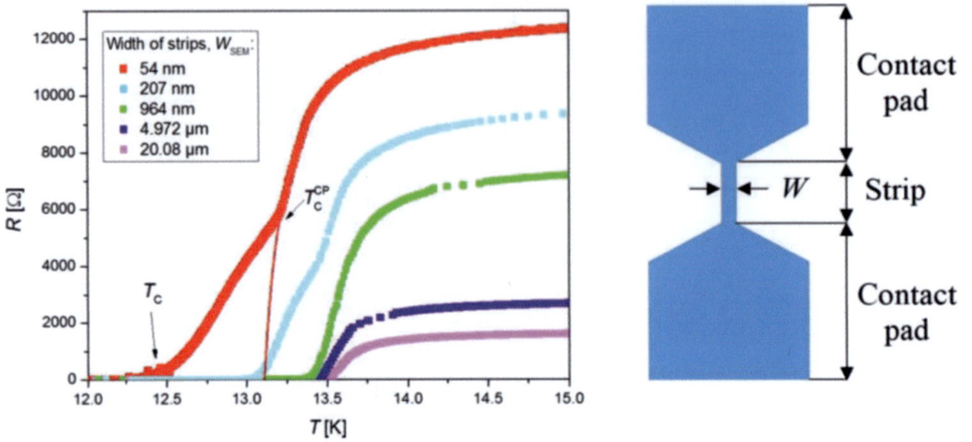

Figure 4.2: Temperature dependence of the resistance of several nano- and microstrips made by negative-PMMA lithography. The width of strips is indicated in legend. The *R(T)* curves of nanostrips have pronounced double-transition which is associated with transition of the contact pads, T_C^{CP}, at higher temperature (fitted by solid line) and transition of the strips, T_C, at lower temperature. The width of contact pads is much larger than nanostrip as shown schematically in the sketch.

The red line is described by 2D fluctuation conductivity theory of Aslamazov and Larkin [81]. The fitting line follows the critical temperature which corresponds to $T_C^{CP} = 13.1$ K of contact pads and area of smooth transition from strip to contact pads as well. Below this temperature, the resistance continuously drops till R starts to fluctuate around 0, as it's cooled far below thought T_C. The critical temperature of a strip was determined by 0.1% of R_{25} criterion, i.e. as a temperature at which the resistance of measured strip drops below $10^{-3}R(25\text{ K})$. For a typical resistance value of strips (~ 10 kOhm) this criterion corresponded to the measured voltages well above the noise level of the experimental setup. The double-stage transition is not observed in $R(T)$ dependence of microstrips since T_C of the strips is comparable with T_C of contact pads and the non-structured film.

The dependencies of the critical temperature on the nominal strip width, W_{SEM}, for P- and N-chips are shown in Fig. 4.3. For strips wider than 1 μm, the critical temperatures of P- and N-strips are close to each other and to the T_C of the non-patterned NbN films and weak dependent on the width. The critical temperature decreases monotonically with decreasing width for both series of strips in the sub-micrometer range. It is seen that at $W_{SEM} < 200$ nm the suppression of T_C is stronger for P-strips so that the critical temperature of N-strips is about 0.5 K higher in the range of width from 50 to about 100 nm.

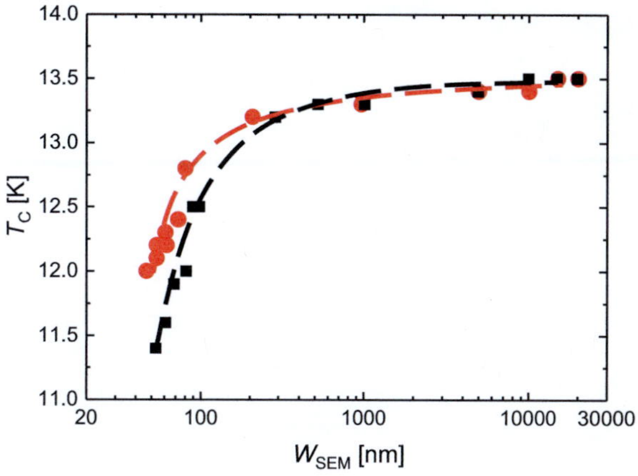

Figure 4.3: Dependence of the critical temperature on the nominal width, W_{SEM}, of positive-PMMA strips (black squares) and negative-PMMA strips (red circles). Dashed lines are to guide the eyes.

The nominal residual resistivity of the strips at $T = 25$ K, ρ_0, was calculated from the resistance with account for the number of squares in the design and the thickness of the film. The dependencies of ρ_0 on the nominal width are shown in Fig. 4.4. The resistivity ρ_0 of both series of strips grows monotonically with decreasing width. The residual resistivity of P- and N-strips with width larger 1 μm are similar to each other and about 10% larger than the resistivity ρ_{25} of non-patterned film. In the range of width below 100 nm, ρ_0 of P-strips is larger than ρ_0 of N-strips. The maximal values of the nominal residual resistivity are 169 and 186 μΩ cm for N- and P-strips, correspondingly.

The second critical magnetic field at $T = 0$ K was estimated from $B_{C2}(T)$ in the dirty limit as $B_{C2}(0) = 18.8$ T. The correspondent value of the coherence length at zero temperature was estimated as $\xi(0) = 4.2$ nm, which is about the thickness of the film. The current-voltage characteristics of all strips were measured at 4.2 K in the current bias mode. The critical current $I_C(4.2\text{ K})$ of a strip is associated with a well-pronounced jump of voltage from zero to a finite value corresponding to the resistive state of the strip. The dependencies of the nominal density of critical current, $j_C(4.2\text{K}) = I_C(4.2\text{K})/(W_{SEM}d)$, on width for both series of strips are shown in Fig. 4.5. In contrast to T_C, $j_C(W_{SEM})$ dependencies are non-monotonic. The j_C increases with increasing width from 50 up to 250 nm, reaches a maximum value at about 250 nm and then decreases with width in micrometer range. Similar dependencies of j_C on width were reported for Nb [82] and TaN [83] strips with different thicknesses. The j_C values of N-strips (red circles) are only slightly higher than j_C of P-strips (black squares) for $W_{SEM} > 1$ μm. However, for widths smaller 1 μm the critical current density of N-strips is about 30% higher (15 ÷ 17 MA/cm^2) than j_C of the P-strips (10 ÷ 14 MA/cm^2).

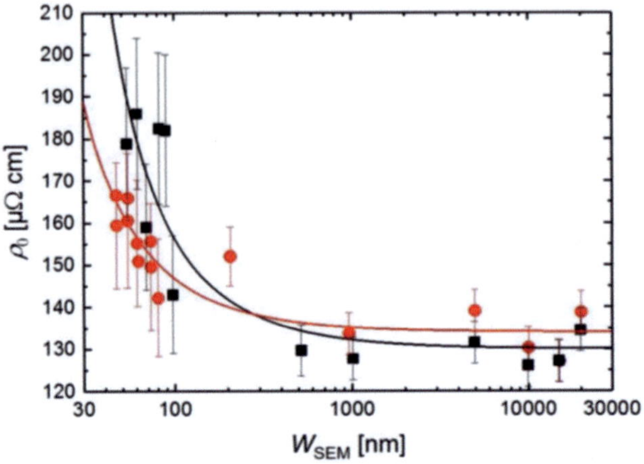

Figure 4.4: Dependence of residual resistivity on nominal width, W_{SEM}, of strips made by positive-PMMA (black squares) and negative-PMMA technology.

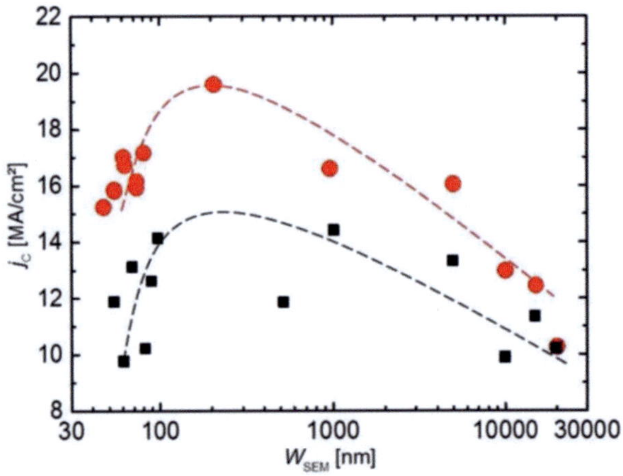

Figure 4.5: Dependence of the nominal density of critical current, $j_c = I_c/dW_{SEM}$, at 4.2 K on nominal width, W_{SEM}, of P-strips (black squares) and N-strips (red circles). Dashed lines are to guide the eyes.

The retrapping current, I_r(4.2 K), is determined as a current at which the strip returns from dependence on width for the micrometer wide strips (Fig. 4.6, black squares) and increase from 2.8 up to 3.1 MA/cm^2 in the range of width smaller than 100 nm. Contrary, j_r of the N-strips increases from 3 up to 3.8 MA/cm^2 with decreasing width (Fig. 4.6, red circles) in all range of nominal width. While the density of the retrapping current of micrometer wide strips is very similar for both N- and P-strips, in case of the strip narrower than 100 nm, the difference in j_r between the strips is about 20%.

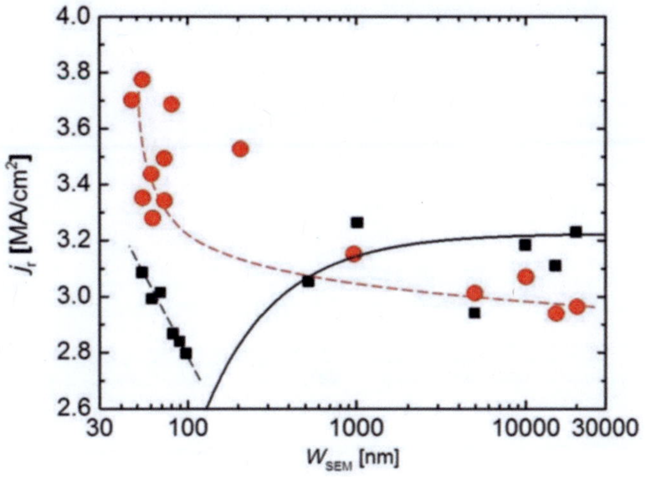

Figure 4.6: Dependence of the nominal density of retrapping current, $j_r = Ir/W_{SEM}d$, at 4.2 K on nominal width, W_{SEM}, of P-strips (black squares) and N-strips (red circles). The dashed lines are to guide the eyes. The solid line is the $j_r^{nom}(W_{SEM}) = \frac{j_c \cdot (W_{SEM} - \Delta W)}{W_{SEM}}$ dependence, where $jr = 3.2$ MA/cm^2 and $\Delta W = 25$ nm.

To investigate the $j_r(T)$ dependence on width of strips, j_r is measured in the temperature range from T_C down to 4.2 K for several nano- and microstrips made with both patterning technics (Fig. 4.7). The j_r grows with reduction of temperature for all measured strips. At lower temperatures, a weak $j_r(T)$-dependence is observed independently of the width of measured strips.

The experimental $j_r(T)$ dependence is described by the equation presented in [84]:

$$j_r = \sqrt{\frac{B}{4d\rho} \times \left(T_C^4 - T_b^4\right)}, \tag{4.1}$$

where B is the thermal coupling coefficient between film and substrate and T_b is the bath temperature. The experimental $j_r(T)$ dependencies presented in graph in Fig. 4.7 are described by (4.1) in the whole temperature range. An example of a well fitted 87 nm narrow strip is shown by pink line in Fig. 4.7. The thermal coupling was estimated to be about 1650 W m^{-2} K^{-4} which is slightly higher than reported in [84].

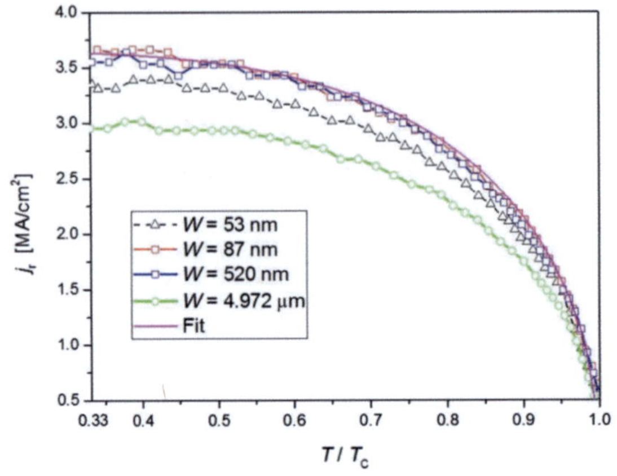

Figure 4.7: Temperature dependence of retrapping current density for several N-strips with different widths (indicated in the legend). The $j_r(T)$ dependence is similar for P- and N-strips in the whole range of widths. The solid lines are the best fits of experimental results by (4.1).

In conclusion, transport measurements of strips made by positive- and negative-PMMA lithography were done using standard probe technique. The measured and estimated superconducting and transport properties of the strips are required for a systematic study of proximity system (N-S-N) in superconducting thin strips which appears naturally during patterning process due to destruction of superconductivity in bands at both edges of the strip.

4.2 Temperature dependence of the critical current

The critical current of particular strip was measured in the temperature range from 4.2 K up to T_C. The dependence of the critical current density on reduced temperature $t = \left[1 - T/T_C\right]^{3/2}$ (the temperature dependence of the Ginzburg-Landau critical current) of narrow nanostrips and wide microstrips are shown for P-strips (Fig. 4.8) and N-strips (Fig. 4.9), respectively.

The shape of the temperature dependence of the critical current density is dependent on strip width. The temperature dependencies of j_C of P- and N- narrow strips with $W < 200$ nm are qualitatively similar to each other (Fig. 4.8a and Fig. 4.9a). With increasing width in this range, the temperature deviation of $j_C(t)$ (Fig. 4.10) from the linear behavior shifts towards higher temperatures closer to T_C and at $W \geq 500$ nm this deviation is hardly recognizable. The deviation occurs at lower relative temperatures, T/T_C, for P-strips (Fig.4.10).

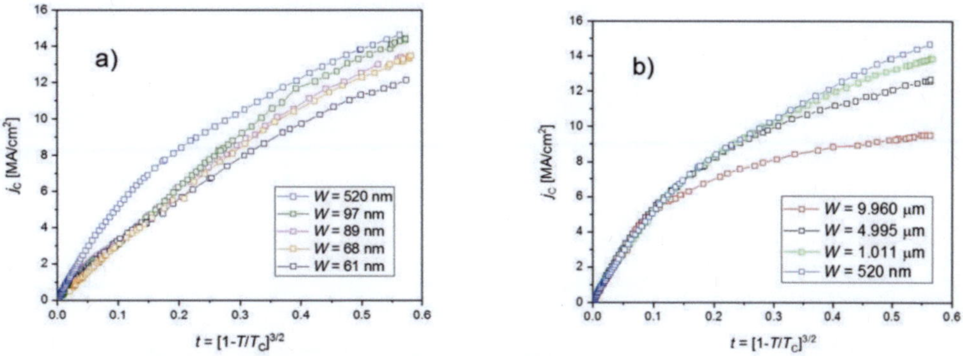

Figure 4.8: Dependencies of the nominal density of critical current on reduced temperature of nano- (a) and microstrips (b) made by positive-PMMA lithography.

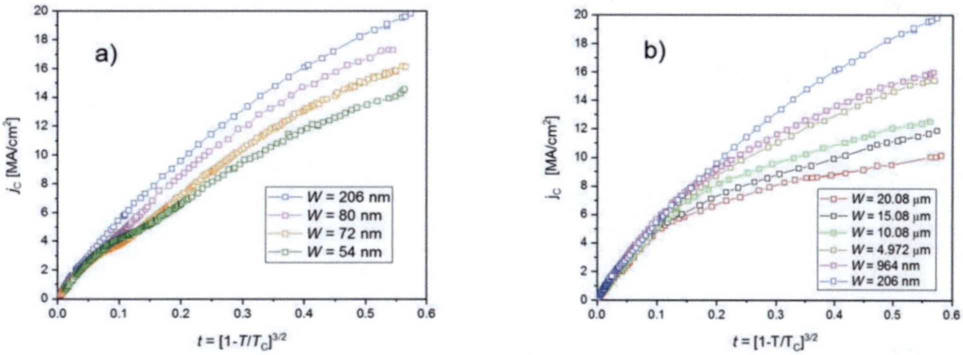

Figure 4.9: Dependencies of the nominal density of critical current on reduced temperature of nano- (a) and microstrips (b) made by negative-PMMA lithography.

The $j_C(t)$-dependence of wide strips is smooth as it is seen in Fig. 4.8b and Fig. 4.9b. The 520 nm strip on Fig. 4.8a has a qualitative different $j_C(t)$-dependence than narrower strips. The specific $j_C(t)$-curves of three strips with different width (54 nm, 207 nm, and 10 μm) are plotted in Fig. 4.11. The curve corresponding to the narrowest strip ($W_{SEM} \approx 54$ nm) looks qualitatively similar to the curve of the 80 nm width strip presented in Fig. 4.9a. The $j_C(t)$-curve of the 207 nm-wide strip coincides with the $j_C(t)$ of the 54 nm wide strip at high temperatures till the deviation from the linear growth of the latter at $t \approx 0.04$. At lower temperatures, j_C of the 207 nm-wide strip increases monotonically and at the lowest measured temperature ($T = 4.2$ K) its value is more than 30% higher than j_C of the 54 nm-wide strip. The third curve in Fig. 4.11 corresponds to the 10 μm-wide NbN strip. The j_C of this strip increases monotonically with decreasing temperature. At temperatures in vicinity of T_C the $I_C(t)$-dependence of the widest strip coincides with the curves of the narrower strips.

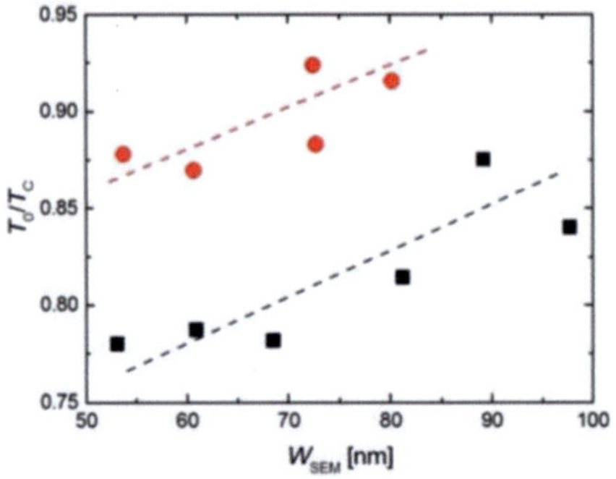

Figure 4.10: Dependence of reduced temperature of the interleaving, T_0/T_C, on nominal width of P- (black squares) and N- (red circles) strips.

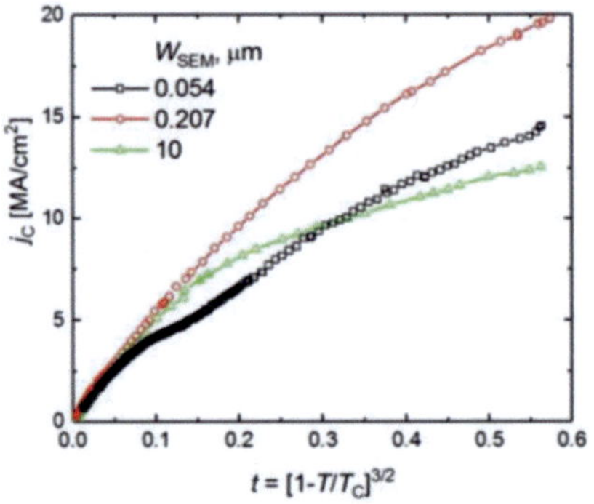

Figure 4.11: Dependencies of nominal density of critical current on reduced temperature of three N-strips of different width indicated in the legend.

Then at temperatures higher t, the j_C of the 10 μm wide strip exceeds the j_C of the narrowest strip and follows the curve of the 207 nm-wide NbN strip till t which is slightly smaller 0.1. Then at lower temperatures (larger t values) the curve of the 10 μm-wide strip deviates from the $j_C(t)$ of 207 nm-wide strip. The growth of j_C becomes shallower and at $t \approx 0.3$ the curve of the 10 μm-wide strip crosses the curve of the 54 nm-wide strip so that at $T = 4.2$ K the j_C of 10 μm-wide strip is about 20% smaller than j_C of the narrowest strip.

The $j_C(t)$-curve of the 207 nm-wide strip is the highest among all other strips on the chip. The increase of width of the strips from about 200 nm leads to a lowering of the $j_C(t)$ curves. The decrease of width below 200 nm results in the appearance of the transition region of the $j_C(t)$-curve from higher (at $t < t_0$) to lower (at $t > t_0$) level and to a shift of the deviation temperature, t_0, towards lower temperatures. The increase of width above 200 nm leads to a deviation of $j_C(t)$-curve from the highest curve (corresponding to the 207 nm-wide strip) and to a shift of temperature of this deviation to higher absolute temperatures. Moreover, the dependence of j_C on temperature becomes shallower with increasing width above 200 nm.

In conclusion, two series of P- and N-strips were characterized by critical-current measurement in the temperature range from 4.2 up to 300 K. It was found experimentally that the shape of the $j_C(t)$-curves depends on strip width. This dependence is observed in P-strip and N-strip as well. In narrow strips, the larger damaged edge-bands result in a stronger reduction of the critical temperature at small widths, smaller densities of the critical current and the retrapping current, and larger residual resistivity. The effect is more pronounced in ultra-narrow strips with widths less than 100 nm.

4.3 N-S-N model

To describe experimental results presented in the previous section, the N-S-N model which is schematically represented in Fig. 4.12 is proposed. It assumes that the material properties of NbN are not uniform across the width but edges at both sides of strip are damaged. The properties of the damaged edges are not known. The damaging is caused by Ar-ion milling, which can lead to a decrease of thickness of parts of strip which was not fully protected during the etching process. The decrease of thickness of the superconducting film results in suppression of superconductivity due to the proximity effect [71]. When this decrease is performed by the milling process then the suppression of superconductivity can be even stronger in comparison to the situation when the thickness is varied during deposition. Especially it is expected when the thickness is in the vicinity of superconductor to isolator transition ($d < 2$ nm in case of NbN). Bombardment of the film by high energy ions during etching process can stimulate this transition already at larger thicknesses. A further mechanism responsible for the suppression of superconductivity in the edges is their oxidation.

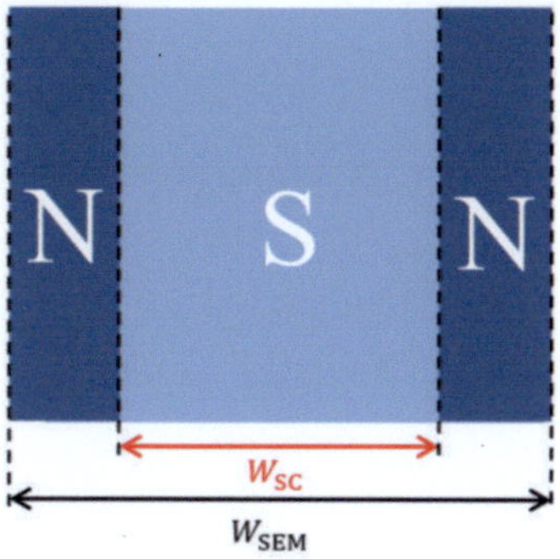

Figure 4.12: A sketch of a lateral proximity system (N-S-N) in a narrow superconducting NbN strip.

After etching, a strip is taken out of the vacuum chamber and exposed to a normal air. Side walls of edges of the strip are not protected by resist; ion milling leads to "activation" of these areas by removing of nitrogen atoms and liberation of Nb bonds attractive to oxygen. Oxides of Nb are either superconducting at temperatures well below 4.2 K or even isolating. The latter reduces the effective conducting thickness of the film further.

The model assumes a uniform suppression of superconductivity over the whole width of edges. In reality, the strength of suppression is distributed from the strongest at the very edge till the weakest at the virtual border between edge and the superconducting central strip part (the superconducting core). The actual profile and the strength of suppression of superconductivity in edges is unknown but could play a crucial role in different phenomena like penetration of vortices, distribution of order parameter and supercurrent density across the width in strip. In frame of the model, the strip is considered as a normal metal–superconductor–normal metal (NSN) structure where the superconducting width, W_{SC}, is smaller than the nominal one, W_{SEM}, by a value ΔW.

For all fabricated strips, especially those with a width smaller than 100 nm, the DC characteristics of N-strips correspond to a "stronger" superconductivity: higher T_C, j_C and j_r values. It assumes that the width of the damaged bands (ΔW) along strip edges is smaller for these N-strips made by the negative-PMMA lithography. This is due to the difference in both the profile and hardness of PMMA resist after exposure and development, which is caused by significant difference in the exposure dose [85]. In other words, a non-damaged width (the effective superconducting width), W_{SC}, is larger in case of strips made by negative-PMMA technology.

Lateral proximity effect in NSN model of NbN strip

The order parameter $\psi^2 \propto \Delta$ in superconducting part of the strip is:

$$\Psi_S = \tanh\left(\frac{x - x_0}{\sqrt{2}\xi_S}\right) \tag{4.2}$$

In the dirty limit $\ell << \xi_S$ the coherence length in the superconductor at zero temperature is

$$\xi_s = \left(\frac{1.781 \cdot \mathrm{h} \cdot v_S \cdot \mathrm{l}_S}{\pi^2 \cdot k_B \cdot T_C}\right)^{\frac{1}{2}} \tag{4.3}$$

and x_0 can be found from the equation

$$-\sinh\left(\sqrt{2}\,\frac{x_0}{\xi_S}\right) = \sqrt{2}\,\frac{\rho_N}{\alpha \cdot \rho_S}\,\frac{\xi_N}{\xi_S} \tag{4.4}$$

where ξ_N is the temperature dependent coherence length in normal metal layer in the dirty limit

$$\xi_N = \left(\frac{\mathrm{h} \cdot v_N \cdot \mathrm{l}_N}{6 \cdot \pi \cdot k_B \cdot T}\right)^{\frac{1}{2}} \tag{4.5}$$

and $\rho_{N,S}$, $\ell_{N,S}$, $v_{N,S}$ are the resistivity, the electron mean free path, and the Fermi velocity in normal metal and superconductor.

In the normal part of strip the order parameter depends on the distance x from the NS-interface as

$$\Psi_N = \Psi_0 e^{-\frac{|x|}{\xi_N}}, \tag{4.6}$$

where Ψ_0 is the value of the order parameter at the interface on the normal side and can be, in general case, lower than the order parameter on the superconducting side.

In case of experimental situation, the barrier between N and S parts could be assumed ideal with equal values of the order parameter at both sides of the interface.

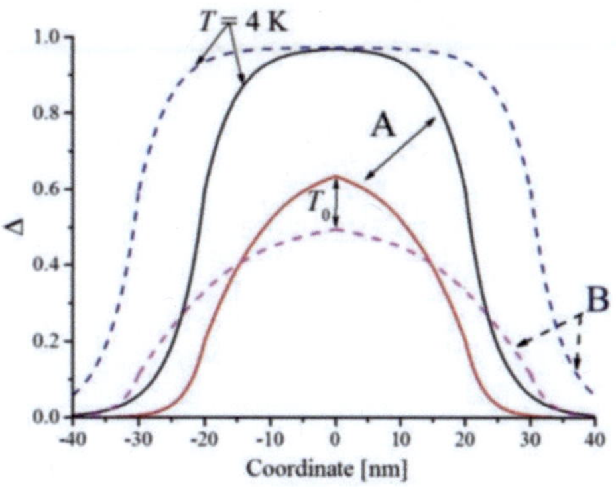

Figure 4.13: Distribution of the gap in two strips of nominal width 80 nm and different values of normal bands 20 nm A-strip (solid lines) and 10 nm B-strip (dashed lines) calculated for $T = 4$ K and at correspondent interleaving temperatures T_0 (chosen for P- and N-strips of correspondent width.

The dirty limit approximation is valid for thin NbN films. The calculated values of the electron mean free path are orders of magnitude smaller than the coherence length obtained from the measurement of the second critical magnetic field, Eq. 4.1. The values of the Fermi velocity and the electron mean free path in the normal part of strip are not known but can be assumed to be not larger than in the superconducting core. Based on these assumptions the distribution of the order parameter across the width of the strip of nominal width 80 nm for two different values of width ΔW of normal edges 20 nm (sample A) and 10 nm (sample B) at both sides is calculated. The calculations were made at two different temperatures: $T1 = 4$ K which is well below T_C and corresponds to the lowest temperature of the measurements; $T2$ was chosen as $0.8T_C$ and $0.9T_C$ for wider (sample A) and narrow (sample B) normal edges, correspondingly. The latter values (corresponding to the interleaving temperature) were taken from Fig. 4.10 for about 80 nm-wide P- and N-strips. The results of calculations are shown in Fig. 9.14. At low temperature, profiles of the order parameter are quite similar for the both samples. The order parameter reaches a certain saturation value in the middle of the strip.

The maximal values of Ψ are close to each other. However, the decay of the order parameter of sample A begins at smaller distance from the center and reaches the minimal value at the very edge of the sample. The value of Ψ of sample B is larger than the order parameter of sample A over the whole nominal width of strip. Areas under Ψ(x)-curve are proportional to the nominally superconducting width.

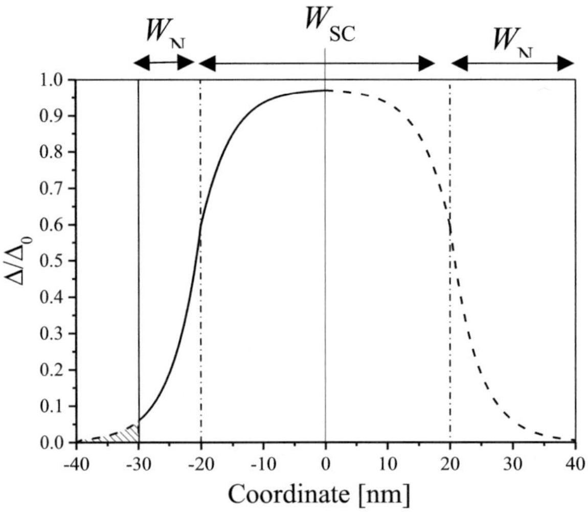

Figure 4.14: Calculated distribution of the gap at $T = 4.2$ K in two strips of the same superconducting width (W_{SC}) and different normal bands $W_N1 = 10$ nm (left) and $W_N2 = 20$ nm (right). The hatched area in the left part of the graph is the difference in the gap distribution between strip with W_N2 and W_N1.

At the interleaving temperatures (*T*2) the profile of the Ψ(x)-curves is considerably different. First, there is no saturation of the order parameter in the center of strips. The maximal value of Ψ of the sample A is more than 20% larger than that one of sample B. However, at the very edges of the strip the ratio of order parameters is opposite: the Ψ value of sample B is an order of magnitude larger than Ψ of sample A. Nevertheless, the areas under the Ψ(x)-curve corresponding to interleaving temperatures are almost equal to each other.

The suppression of the critical temperature is stronger for P-strips which are with a smaller superconducting width at the same nominal width. However, if the superconducting width is the same for strips with different width of damaged edges (Fig. 4.14) the Ψ(x) distribution inside the superconducting core does not depend on width of normal edges. The difference in distribution of the order parameter is at the very edges only (marked in Fig. 4.14 by the hatched area) and its contribution to the total superconducting strength is negligible and thus not measurable in ordinary experiments. A more precise and quantitative analysis and comparison of experimental data with theoretical calculations requires exact data of experimentally determined distributions of the order parameter across the width with a spatial resolution of at least about the coherence length.

4.4 Magnetic-field dependence of the critical current density

Measurements of the critical current in magnetic fields B have been done at $T = 6.2$ K in magnetic fields up to 1 T applied normally to the sample surface. The dependence of the critical current density on magnetic field for nano- and microstrips is shown in Fig. 4.15. At small fields, the dependence of j_C on B has a sharp maximum and is fully symmetric with respect to zero magnetic field which is expected for superconducting structures without any defects [56]. The so-called B_{stop} is defined by the crossover between Meissner and mixed states in the strip. At fields higher than B_{stop}, the dependence becomes weaker. The mixed state is stable until the current becomes large enough for vortices to move and the strip enters the dynamic mixed state. This current defines the critical current for the high magnetic field range and follows a dependence of critical current in magnetic field ~$1/B$.

According to [66], B_{stop}, which is magnetic field corresponding to the first vortex penetration into the superconducting strip of width W, is estimated as:

$$B_{stop} = \frac{\Phi_0}{2\sqrt{3}\pi\xi(T)}\frac{1}{W} \qquad (4.7)$$

where $\xi(T)$ is the temperature dependent value of the coherence length

$$\xi(T)=\xi(0)\left[1-\frac{T}{T_C}\right]^{-\frac{1}{2}}\left[1+\frac{T}{T_C}\right]^{-\frac{1}{4}} \tag{4.8}$$

and $\xi(0)$ is determined by Eq. 4.1.

The Meissner state in the strip is described as:

$$j_C(B)=j_C(0)\left(1-\frac{B}{2B_{\mathrm{stop}}}\right) \tag{4.9}$$

where $j_C(0)$ is the critical current at $B = 0$ T. The fit of the $j_C(B)$-dependence of several NbN strips by Eq. 4.9 is shown by solid lines in Fig. 4.15. B_{stop} decreases with increasing strips width.

The value of B_{stop}, the transition of the strip from the Meissner to vortex state, which corresponds to deviation of $j_C(B)$-curve from the linear decrease, is determined as the only fit parameter in Eq. 4.9.

The extracted B_{stop} from fitting for the whole range of width is plotted as a function of the W_{SEM} on Fig. 4.16. For widths wider than 1 μm, B_{stop} for P- and N-strips coincides with the theoretical curve calculated in frame of Ginzburg-Landau model [86].

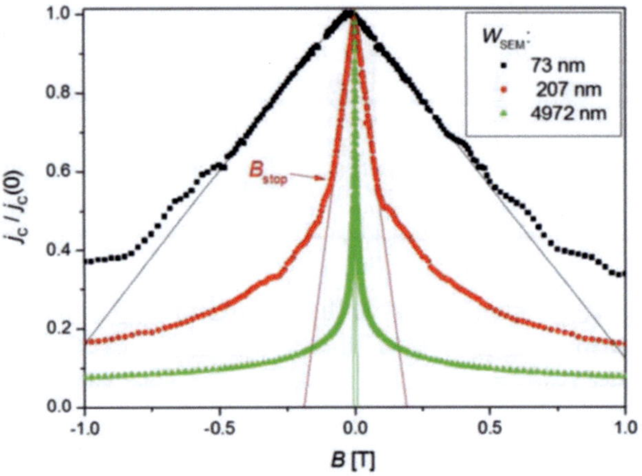

Figure 4.15: $j_C(B)$- dependence of P-strips measured at 6.2 K. The width of strips is indicated in legend.

The deviation of the extracted B_{stop} from theoretical curve is found at $W_{SEM} < 1$ µm. Physically, a higher B_{stop} means smaller superconducting width in the strip. The difference between experimentally measured and theoretically predicted values of B_{stop} is larger for strips with expected larger width of normal bands made by positive-PMMA lithography.

It was also found that defects in strips provide asymmetry in the $I_C(B)$ dependence. During measurements of I_C in the magnetic field the asymmetrical behavior of $I_C(B)$ is observed.

The asymmetry is associated with current crowding near the defect at the edge of the strip. The corresponding asymmetrical $I_C(B)$ dependence is presented in Fig. 4.17a. Due to suppressed superconductivity at place with defect, the supercurrent flows along the periphery of the intrinsic defect. The current is locally redistributed and the density of the current increases in the vicinity of defects. By applying an external magnetic field, the local density of the super-current further increases and a corresponding reduction of I_C is observed. In the opposite field direction, the density of critical current in vicinity of defect decreases. This leads to an enhancement of supercurrent of the strip at small magnetic fields.

The maxima of I_C at 0 T flattens in small range of field when the local defect in film is inside the strip. Instead of symmetrical $I_C(B)$ dependence (shown by blue solid lines on Fig. 4.18b), depleted I_C was measured at magnetic field ±20 mT for a given sample.

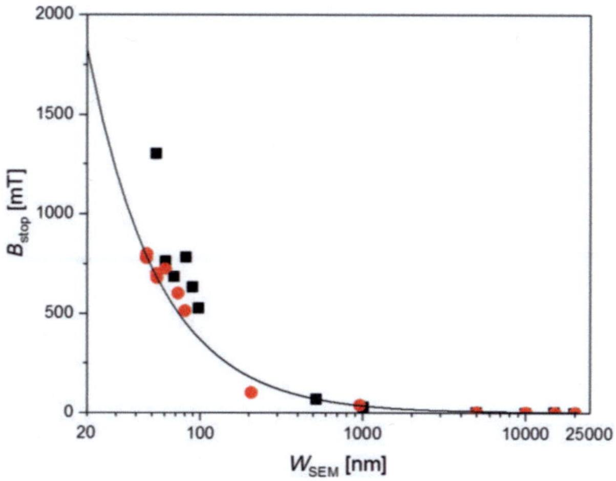

Figure 4.16: Dependence of magnetic field corresponding to the penetration of the first vortex, B_{stop} (symbols), on W_{SEM} of strips made by positive- (black) and negative-PMMA (red) lithography. The theoretical $B_{stop}(W_{SEM})$ dependence is calculated by Eq.4.7 and shown by solid line.

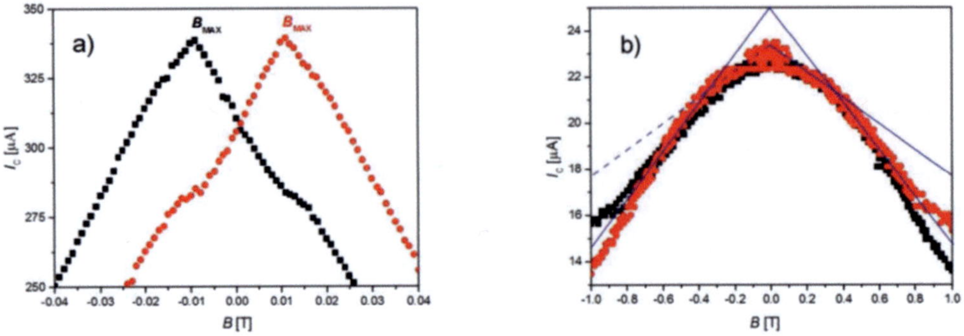

Figure 4.17: $I_C(B)$-dependences of strips with defects at the edge (a) and inside of strip (b).

4.4.1 Superconducting width in nanowires

While the nominal width W_{SEM} can be measured by using SEM inspection, evaluation of the effective superconducting width requires measured values of B_{stop}. The B_{stop} field is obtained from the measurements of dependence of critical current density on magnetic field (described in previous section). The effective superconducting widths of P- and N-strips, which were calculated out of the experimentally determined B_{stop} values using Eq. 4.7 (B_{stop}), are shown in Fig. 4.18 in dependence on the nominal width W_{SEM}. It is seen that W_{SC} of almost all strips is smaller than the nominal width W_{SEM} in the whole range of studied geometries and the used patterning technologies. In turn, the effective superconducting width of the NbN strip made by negative-PMMA technology is larger than W_{SC} of the strips made by positive-PMMA technology.

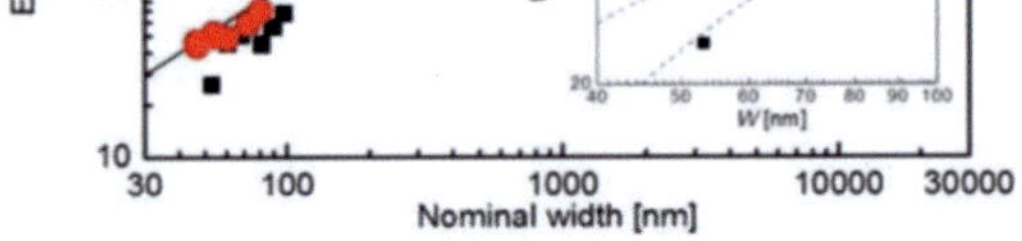

Figure 4.18: Dependence of the effective superconducting width, W_{SC}, calculated by Eq. 4.7 on the nominal width, W_{SEM}, of the P- (black squares) and N-strips (red circles). The solid line corresponds to $W_{SC} = W_{SEM}$. The insert is a close up of the dependence for $W_{SEM} < 100$ nm. The dashed lines are to guide the eyes.

The dependence of calculated residual resistivity on nominal width of strips made by both technologies is shown in Fig. 4.4. Qualitatively both experimental curves look similar to each other: the resistivity is almost independent of width in the range above ≈ 1 μm and increases with decreasing width below this limit. These dependencies were analyzed in frame of the lateral NSN model of NbN strip. The strip with damaged edges is considered as a system of two resistors with different resistances connected in parallel. The resistance of a central part of the strip of width $W_{SC} = W_{SEM} - \Delta W$ (W_{SEM} is the nominal width of strip and ΔW is the total width of two damaged edges) is $R_{SC} = \dfrac{\rho_{SC}}{W_{SC}}$ where ρ_{SC} is the resistivity NbN film in this part which is assumed to be lower than the resistivity of normal (damaged) edges $\rho_N = \rho_{SC} + \Delta\rho$. The effective resistivity of the strip can be calculated as

$$\rho_{\mathrm{eff}}\left(W_{\mathrm{SEM}}\right) = \frac{\rho_{\mathrm{SC}}\left(\rho_{\mathrm{SC}} + \Delta\rho\right) W_{\mathrm{SEM}}}{\rho_{\mathrm{SC}} W_{\mathrm{SEM}} + \Delta\rho\left(W_{\mathrm{SEM}} - \Delta W\right)} \tag{4.10}$$

At $W_{\mathrm{SEM}} >> \Delta W$ the effective resistivity of the strip, ρ_{eff}, is independent of the nominal width and equal to ρ_{SC}. According to the results presented in Fig. 4.18, the width of damaged edges is not larger than 40 nm. Therefore, the saturated value of resistivity for $W_{\mathrm{SEM}} > 1$ μm can be assigned to ρ_{SC}, solely. The main influence on ρ_{eff} from the damaged edges is expected when their width, ΔW, is comparable to W_{SEM}. Indeed, the strongest increase of resistivity is observed for $W_{\mathrm{SEM}} < 100$ nm (Fig. 4.4). The ΔW value varies from 5 to 10 nm in case of N-strips and from 15 to 25 nm in case of P-strips. It's assumed that the strength of damage of NbN film at edges and the correspondent increase of resistivity ($\Delta\rho$) are independent of the used technologies. Based on the above-mentioned assumptions both $\rho_0(W_{\mathrm{SEM}})$ dependences are fitted by Eq. 4.10 (the solid lines in Fig. 4.4). The best fit was obtained for $\Delta\rho$ about 160 μΩ cm. This means that the resistivity of damaged edges is about twice larger than the resistivity of the superconducting core of the strip, $\rho_{\mathrm{SC}} \approx 130$ μΩ cm. In spite of the variation of ΔW for the fitting, the values 30 and 16 nm for P- and N-strips are used, which are very close to the correspondent average values of ΔW: 25 and 10 nm (shown by the dashed lines in the inset in Fig. 4.18). In Fig. 4.4, it's seen that the fit describes the data of N-strips pretty good while the P-strips show steeper increase of the residual resistivity. The discrepancy between the fit and the experimentally obtained resistivity of the positive-PMMA strips can be explained by the simplicity of model (Eq. 4.11) which does not account, for example, for contributions to the total resistance of the wiring pads and of the rounded transitions from these pads to the strips as well as possible dependence of $\Delta\rho$ on W_{SEM}.

In Fig. 4.19, the critical temperature is plotted versus the experimentally determined W_{SC}. The experimental $T_C(W_{SC})$-dependencies of both series of NbN strips can be described now by one universal curve in the whole range of the effective superconducting width. This is in agreement with the predictions which are made in frame of the NSN model (Fig. 4.14): the strength of suppression of T_C due to the lateral proximity effect is weakly dependent on the width of damaged edges of highly disordered NbN strips. The solid line in Fig. 4.19 is the dependence of critical temperature of NS proximity structure [87]

$$T_C = T_{C0}\left(1-\left(\frac{\pi\xi_{GL}}{W_{SC}}\right)^2\right) \tag{4.11}$$

obtained in case of a normal layer resistivity is much lower than resistivity of the superconducting layer. The $T_{C0} = 13.48$ K and $\xi_{GL} = 4.2$ nm have been used for the fit.

An increasing resistivity of the normal layer shifts the T_C decay to smaller values of W_{SC}. The result of calculations made as in [88] for the ratio $\rho_{SC}/\rho_N = 2$ is shown in Fig. 4.19 by green triangles. Both theoretical calculations describe the experimental results qualitatively well. However, the theory predicts a weaker suppression of T_C for strips in which the resistivity of the edges is larger than the resistivity of the superconducting core.

In frame of the formalism presented in [88], it was obtained that the suppression of T_C becomes independent of width of normal layer if ΔW is larger than about $1.6\xi_{GL}$. In case of NbN films this limit corresponds to about 7 nm which is about or smaller than the width of damaged edges in the strips. Therefore, the experimentally observed independence of $T_C(W_{SC})$ (Fig. 4.19) on width of the normal edges agrees well with theoretical predictions.

The effective density of retrapping current (Fig. 4.20) is almost independent of width at $W_{SC} \geq 200$ nm and increases with decreasing width below this values from 3.75 MA/cm^2 up to about 4.75 MA/cm^2, i.e. by a factor more than two. The increase of j_r^{eff} in the submicrometer range of width can be explained by an improvement of the cooling efficiency of the NbN strips with width which becomes comparable and smaller than the thermal healing length [89]. Indeed, following [89], the healing length was estimated for structures about 95 nm which is comparable to the starting width of j_r^{eff} growth. For comparison, the thermal healing length was found around 100 nm for 7 nm thick NbN nanowires deposited on a MgO substrate [90]. Other possible mechanisms, which could be a reason of the increase of j_r with decreasing width, is an increase of the ratio of surface S, through which the heat is transferred from strip to environment, to volume V, in which this heat is generated,

$$\frac{S}{V} = \frac{2d+2W}{Wd} \quad . \tag{4.12}$$

This ratio is constant at $W >> d$ and increases as W^{-1} when the width becomes comparable to the thickness d of film.

The experimentally measured nominal density of the retrapping current of N-strips is about 20% larger than j_r of P-strip for $W_{SEM} < 100$ nm (Fig. 4.6). The density of the retrapping current is dependent on specific resistivity ρ_0 of the superconductor in the normal state and on the difference between T_C and bias temperature, T_b, at which a current-voltage characteristic is measured using Eq. 4.1.

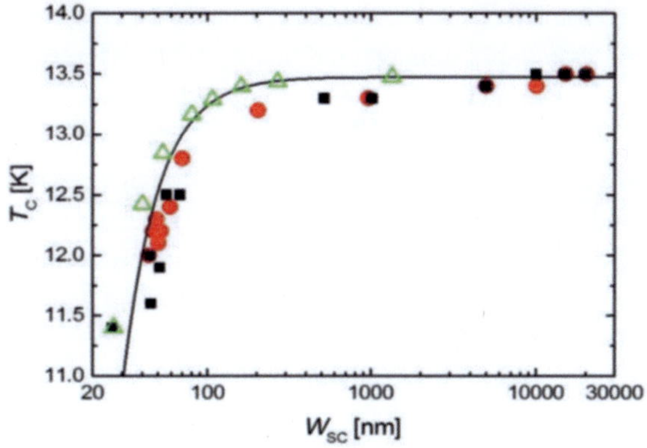

Figure 4.19: Critical temperature as a function of the effective superconducting width. The plotted data is fitted by Eq. 4.11. Green dots represent result of calculations in frame of formalism presented in [88].

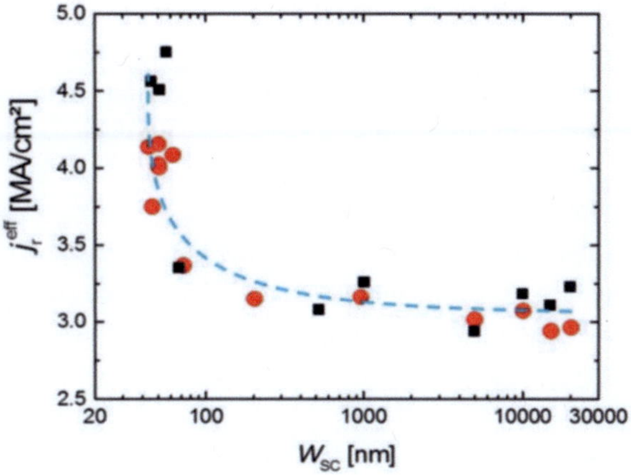

Figure 4.20: Dependencies of retrapping $j_r^{eff} = I_r/dW_{SC}$ on effective superconducting width of strip, W_{SC}. The dashed blue line is to guide the eyes.

In Eq. 4.1, B is the coefficient characterizing the thermal conductivity of the interface between superconducting film and substrate independently of the geometry of a measured strip. Taking larger T_C values into account (Fig. 4.3) and smaller ρ_0 (Fig. 4.4) of N-strips in sub-100 nm range of width the calculated by Eq. 4.14 the difference in the retrapping current densities of N- and P-strips is about 9% only.

In frame of the NSN model this difference between the estimations and experimentally determined nominal densities of j_r can be explained qualitatively in a way which is similar to the one used above to describe the correspondent difference in the nominal densities of the critical current.

The strip is considered as parallel connected resistors with different resistances which are characterized by different temperature dependencies in the range between T_b and T_C also. In case of the damaged edges the resistivity is higher, superconductivity is suppressed and thereby it is natural to assume that in this temperature range the resistivity of the edges can be considered independent of temperature (the fully suppressed superconductivity) or at least the resistivity of edges is higher than resistivity of the superconducting core at a given temperature. This difference in resistivity results in a non-uniform distribution of applied current even in the fully normal state of strip at T well above T_C. It can be shown that a part of the applied current which flows through the superconducting core of strip I_{SC} is

$$\frac{I_{SC}}{I_b} = \frac{1}{\dfrac{\rho_{SC} \cdot \Delta W}{\rho_N \cdot W_{SC}} + 1} \tag{4.13}$$

A decrease of the applied current, in general, leads to a decrease of generated Joule heat power that at permanent cooling conditions (assumed to be independent of I_b) subsequently decreases the effective temperature of the strip and T-dependent parameters ρ_{SC} and ρ_N as well. Since ρ_{SC} decreases with temperature stronger than ρ_N then according to Eq. 4.13 the part of bias current through the superconducting core increases. The correspondent decrease of current through the damaged edges results in even stronger decrease of contribution of the edges into generation of Joule heat. At low enough current, which is close to the retrapping current, the largest part (if not all) of the applied bias current flows through the superconducting core excluding the damaged edges almost completely. In this situation the retrapping current density is determined by the thermal interface between the core and substrate, T_C and ρ_{SC} solely, i.e. j_r should be independent of a value of $W_{SC} = W_{SEM} - \Delta W$. Thereby the nominal density of the retrapping current

$$j_r^{nom} = \frac{j_r \cdot W_{SC}}{W_{SEM}} \tag{4.14}$$

is proportional to the value of width of the superconducting core, i.e. j_r^{nom} is larger for N-strips in agreement with experimental observations. However, according to Eq. 4.14 the nominal density of retrapping current should decrease with the decreasing nominal width (the solid line in Fig. 4.6) if reasonable to assume that the width of the damaged edges, ΔW, is independent of W_{SEM}. However, the experimentally obtained nominal density of the retrapping currents of P-strips increases with the decreasing W_{SEM} in sub-100 nm range of width and in case of N-strips the increase of j_r is seen in the whole range of studied widths.

Similar to j_r, the situation is seen in Fig. 4.21 where the effective densities of the critical currents ($j_C^{eff} = I_C/(W_{SC}d)$) are plotted in dependence on W_{SC}. The experimental data of j_C^{eff} of both series of the NbN strips can be now described by universal curves. The effective density of the critical current increases in all range of width (Fig. 4.21) from 10 up to 25 MA/cm^2. The dashed horizontal line in Fig. 4.21 corresponds to the value of the depairing current density, which was computed in frame of Ginzburg-Landau formalism with correction for dirty limit, T_C and ρ_0 values for non-patterned film. The deparing current density which was calculated for NbN film is plotted as black dashed line on Fig. 4.21. It is seen that the experimental points approach the theoretical limit of the current carrying ability of the NbN film remain below this limit.

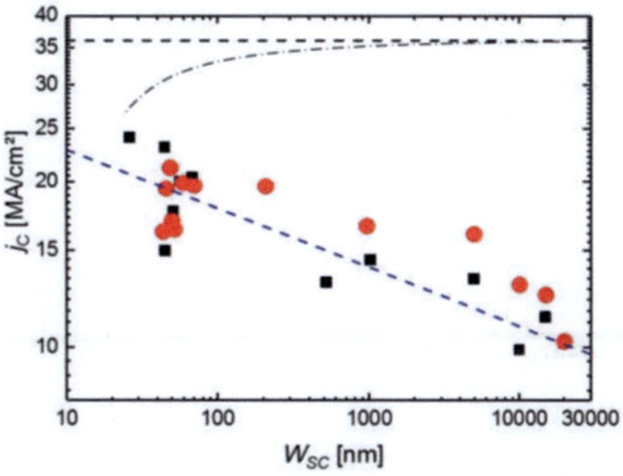

Figure 4.21: Dependencies of $jc = Ic/dW_{SC}$ on effective superconducting width of strip, W_{SC}. Dashed blue line is common curve which describes dependencies of superconducting parameters for both types of strips. Dashed black line on graph corresponds the value of the depairing current density, which was computed in the framework of the standard Ginzburg-Landau (GL) approach with the correction for the extreme dirty limit. The dependence of the depairing current density on the superconducting width calculated with corresponding superconducting properties of strips is plotted as dash-dotted line on the graph.

The temperature dependence of j_C is valid for all strips with width smaller 200 nm. The $j_C(t)$ dependencies of wider strips are characterized by the flattening of experimental curves. This flattening with a trend to saturation becomes stronger with increasing width and is seen as deviation of the $j_C(t)$ of 10 μm-wide strip from the curve corresponding to the 207 nm-width strip (Fig. 4.11). With increasing width above 200 nm the temperature of this deviation increases and the curves below this temperature become flatter so that the difference between the highest reachable j_C(4.2 K) of the 207 nm-wide strip and j_C(4.2 K) of strips in the micrometer range of widths increases (Fig. 4.21).

The mechanisms of this deviation and of the flattening of the temperature dependencies of critical current can be associated with penetration of vortices in the strip [86]. The magnetic field generated by transport current at the edge of the strip is calculated as [91]:

$$B_{edge} = \frac{\mu_0 \cdot I \cdot \ln\left(\frac{2W}{d}\right)}{2\pi \cdot W} \tag{4.15}$$

Using this formula, the corresponding self-field is computed for whole range of strips and compared with value of B_{stop} obtained from $I_C(B)$-dependences. Fig.4. 22 represents results of calculations for both series. The self-field B_{edge} generated by the bias current is much lower than B_{stop}, especially for submicrometer strips. The B_{edge} of P- (open black symbols) and N- strips (open red symbols) smaller 1 μm is three order magnitude lower than the field required for penetration of vortex (solid black line). Although the value of magnetic field required for penetration of vortex decreases with increase of the width of strip, B_{edge} is lower than B_{stop} even for wide strips.

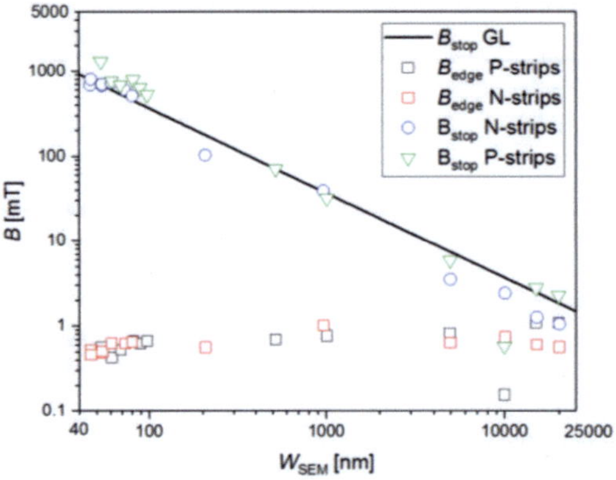

Figure 4.22: Comparison of extracted B_{stop} and calculated B_{edge} for P- and N-strips (indicated in legend). The black solid line represents B_{stop} calculated in frame of Ginzburg-Landau formalism.

4.5 Depairing critical current

The width of all strips is smaller than the Pearl length. The supercurrent is expected to be uniformly distributed over the cross-section of strip and determined by the de-paring mechanism. However, temperature dependencies of the critical current of the strips with $W \leq 200$ nm are non-monotonic in spite of the fact that in this range of width $W << \Lambda$. Therefore, it is not possible to describe the $I_C(T)$-dependencies of these strips by the temperature dependence of the de-paring current in the whole temperature range. The temperature dependence of the Ginzburg-Landau critical current is valid at $T_C - T << T_C$ only. In the dirty limit, $1 << \xi_0$, it can be extended over the whole temperature range with account for the temperature dependent correction, $KL(T)$, found in Fig. 1 in Ref. [92].

The thin solid line in Fig. 4.23 is the linear fit of the critical current at $T \rightarrow T_C$, by $I_C^{extr}(t) = I_C^{extr}(0)t$, where $t = \left(1 - T/T_C\right)^{3/2}$ is the temperature dependence of the Ginzburg-Landau critical current. At temperatures below the interleaving temperature the experimental points deviate from the $I_C^{extr}(t)$-curve. The dashed line in Fig. 4.23 shows the temperature dependence of critical current which was calculated as $I_C1(t) = I^{extr} \cdot KL(t) \cdot t$.

However, even with account for the KL-corrections the measured critical current is below the calculated curve at the reduced temperatures above interleaving temperature. This low-T part of $I_C(t)$ can be described (the thick solid line) up to the maximal $t \approx 0.6$ ($T = 4.2$ K) by assuming that in this temperature range, i.e. far from the critical temperature, the applied supercurrent is concentrated in the central (superconducting) part of the strip while the normal edges with suppressed superconductivity are excluded from the transport of current completely. However, it is seen that at $t \geq 0.1$ the experimental points go parallel to the $I_C1(t)$-curve. This low temperature part of the experimental curve can be described up to the largest $t \approx 0.6$ ($T = 4.2$ K the lowest temperature of the measurements) in frame of NSN model assuming that supercurrent in this t-range is carried by the effective superconducting width of the strip while the damaged edges are excluded from the transport completely. The thick solid curve in Fig. 4.23 is calculated on base of this assumption as $I_C2(t) = I_C1(t)W_{SC}/W_{SEM}$ where the correspondent $W_{SC} = 67$ nm (see the inset in Fig. 4.18).

There are several features in the shown $I_C(t)$ dependence which is typical for strips with width $W \leq 200$ nm (see also the black curve in Fig. 4.11). First of all, in vicinity of T_C ($t \rightarrow 0$) the critical current is linearly proportional to t with a coefficient I^{extr} (the thin solid line in Fig. 4.23). Secondly, the experimental curve deviates from this linear growth at $t_0 \approx 0.04$ ($T/T_C \approx 0.9$), which is marked by the up-arrow in the graph. (in the following this reduced temperature t_0 and correspondent absolute, T_0, and relative temperatures, T_0/T_C, is named the interleaving temperature.)

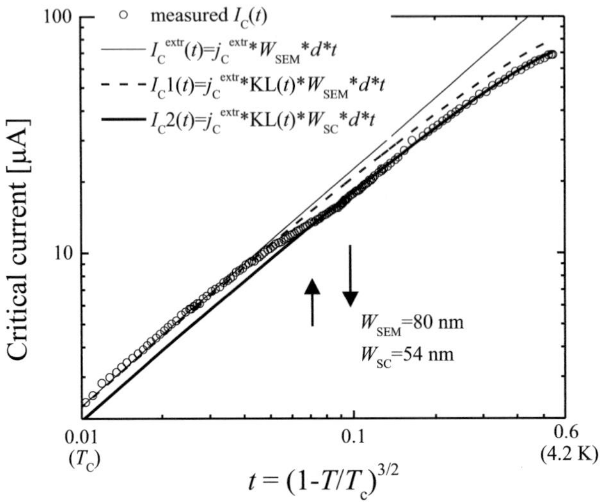

Figure 4.23: Dependence of the critical current (circles) of a 80 nm nanowire (nominal width, W_{SEM}) on reduced temperature $t=(1-T/T_C)^{3/2}$, $I_C(t)$. The thin solid line is the extracted current $I_C^{extr}=j_C^{extr}*W_{SEM}*d$ (the linear fit of the high temperature part $T\rightarrow T_C$ of $I_C(t)$ curve). The dashed line ($I_C1(t)$) is the extracted current multiplied by the temperature dependent correction of Kupriyanov and Lukichev, $I_C^{extr}*KL(t)$. The thick solid line ($I_C2(t)$) is the calculated critical current of the superconducting part (W_{SC}) of the nanowire, $j_C^{extr}*KL(t)*W_{SC}*d$. The arrow marks position of the interleaving temperature T_0.

Thirdly, there is a pretty narrow temperature range at $t \geq t_0$ of transition of the experimental $I_C(t)$-curve to the second, a low-temperature part of the temperature dependence of the critical current. The correspondent $t \approx 0.08$ ($T/T_C \approx 0.8$) of the end of the transition region is marked in Fig. 4.24 by the down arrow.

In order to compare the experimental j_C and theoretical depairing current density in P- and N-strips, the Ginzburg-Landau (GL) depairing current density is computed with the correction $KL(T)$ for the extreme dirty limit as:

$$j_C^{dep}(T) = \frac{16\sqrt{\pi}\exp(2\gamma)}{21\varsigma(3)\sqrt{6}} \beta_0^2 \frac{(k_B T_C)^{\frac{3}{2}}}{e\rho_0\sqrt{Dh}} \left(1-\frac{T}{T_C}\right)^{\frac{3}{2}} KL(T) \quad . \tag{4.16}$$

$$KL(T) \approx 1.87 - \frac{1.46}{1+\left(0.7\frac{T}{T_C}\right)^{1.12}} \tag{4.17}$$

is the analytical approximation of the curve from Fig. 1 in Ref. [92] corresponding to the limit $\xi_{GL} >> \ell$. In Eq. (4.16) e, $\hbar$, k_B are the fundamental constants, $\gamma = 0.577$, $\zeta(3) = 1.202$, $\beta_0 = \Delta(0)/(k_B T_C) = 2.05$ [74] and $D = 0.56$ cm^2/s is the electron diffusivity in NbN film (Eq. 3.3).

Accounting for almost all material parameters having an influence on j_C (Fig. 4.5), i.e. T_C (Fig. 4.3) and ρ (Fig. 4.4) and assuming that the diffusion coefficient, D, should not affect the depairing current too much due to not to strong variation of D between the core and the edges (expected) and its square root presence in Eq. 4.16. The calculated ratio of j_C of P- to N-strips in sub-100 nm range is $\left. j_C^P / j_C^N \right|_{cacl} = 0.89$. In turn in the experiment (Fig. 4.5), a smaller value of the ratio is found to be $\left. j_C^P / j_C^N \right|_{exp} \approx 0.7$.

In frame of NSN model the supercurrent in vicinity of the measured critical current value is concentrated in the superconducting part of the strip. This is true even in case when the edges are not fully normal conductive. If it accounts for non-zero values of the order parameter in the edges the following qualitative model can be considered. At low temperatures ($T << T_C$) a small applied current $I << I_C$ (smaller than the critical current corresponding to the gap value, Δ', in the edges which is smaller than the gap Δ in the center of strip, $\Delta' < \Delta$) is distributed uniformly across the width. Once the density of applied current increases over the critical value corresponding to the energy gap Δ' the edges switch into normal state and the current is redistributed into the central superconducting part of strip with a larger gap value Δ. Then the current can be increased further till the density of current, which is concentrated in the superconducting core, reaches the critical value which is determined by the Δ value. Smaller W_{SC} results in lower measured critical current and thus to a lower calculated nominal density of critical current in strips, $j_C^{nom} = j_C W_{SC}/W_{SEM}$. Because of smaller W_{SC} in P- strips, the nominal j_C in these strips is lower than in N-strips, especially in sub-micrometer range of width that is in the agreement with experimental results shown in Fig. 4.5.

4.6 Summary

In this chapter, a detailed and systematic investigation of the proximity effect is presented for NbN thin film strips with three order of magnitude range of width. The strips were prepared by two technological techniques described in Chapter 3.

The usage of negative-PMMA lithography leads to an enhancement of both superconductivity in strips and their cooling efficiency to substrate. The two series of strips were characterized in the temperature range from 4.2 up 300K. The critical temperature of narrow N-strips is about 5% higher than T_C of P-strips. The improved cooling efficiency of the N-strips resulted in 20% higher values of the retrapping current density. The critical current density of N-strips is found to be up to 40% higher than j_C of P-strips. The variation of properties caused by the difference in technology of patterning is self-consistently explained in frame of the lateral proximity effect model: cross-section of strip is considered as NSN structure consisting of the superconducting core embedded between two normal conducting bands (the damaged edges). The magnetic-field measurements of the critical current were done for all widths. A method for estimation of superconducting width in nanowire from $I_C(B)$-dependence is proposed. Using the obtained superconducting width, the dependencies of T_C, j_C, and j_r on W_{SC} are described universally independently of particular width of the damaged edges.

In the $I_C(T)$-dependence, the local density of the critical current in the superconducting core of narrow strips approaches the density of the Ginzburg-Landau depairing current. The strips with $W > 200$ nm demonstrate flattening of $I_C(T)$ at low temperatures. The flattening is stronger and begins at higher temperatures (closer to T_C) for wider strips. The critical currents are too low to generate magnetic field above the penetration field, B_{stop}. The mechanisms of this deviation and of the flattening of the temperature dependencies of critical current are not fully understood yet.

5 Single-spiral SNSPD

The single-spiral nanowire is considered as a promising alternative to initial meandered nanowire form of detector for broadening spectral bandwidth of SNSPD. According to considerations presented in section 2.2.1, the bend-free nanowires should have a critical current higher than bended structures. Although it has been commonly agreed that the cut-off wavelength, λ_c, increases with the bias current I_b, the exact relationship between I_b and λ_c remains controversial (section 2.2.2). According to the hard-core normal-spot model $1\text{-}I_b/I_C^d \propto \lambda_c^{-1/2}$, while the hot-spot model predicts a linear relationship $1\text{-}I_b/I_C^d \propto \lambda_c^{-1}$. The numerical model invoking Ginzburg-Landau approach to the evolution of the order parameter results in a $1\text{-}I_b/I_C^d \propto f(\lambda_c)$ dependence which falls between the two analytical relations above (see Fig. 2.1 in section 2.2.1). The major fundamental limitation on I_C in meanders is imposed by the current-crowding effect in the turns of the meander. The double-spiral layout was proposed as alternative design for single-photon detectors. In comparison to meanders, detectors with the double-spiral layout demonstrated larger the critical current, I_C, cut-off wavelength, λ_C, and detection efficiency. Although the performance of photon detectors with the double-spiral geometry becomes better, the S- shape bend in the current path through the center of the double-spiral layout still limits the experimental critical current of the specimen. The practical disadvantage of layouts with alternating straight wires and bends is their strong sensitivity to light polarization. The extinction factor increases with the increase in the wavelength [23]. The absorption of spiral detectors is independent on polarization.

In this chapter, the experimental setup and the equipment used to perform the characterization measurements of superconducting nanowire single-photon detectors are introduced. The detection ability for single photons of a bend-free layout [CSD+17] in the form of a single spiral is studied in a wide range of wavelength. The performance of detectors in form of meander, double spiral and single spiral are compared in terms of critical current, detection efficiency, cut-off wavelength and timing jitter. The technology of samples is described in detail in Chapter 3. The experimental results are analyzed and compared with predictions of the theoretical models.

5.1 Cryogenic setup for characterization of SNSPD

Measurements of optical response are performed in a dipstick cryostat which is immersed in a standard ^{4}He-transport Dewar. The cryogenic setup is very similar to setup which was described in [93]. The detectors are kept at an ambient temperature of 4.2 K. They are fixed on a sample holder together with temperature sensors and a bias tee and manually bonded with indium wires to circuit board on the holder. The cooling power at the copper sample holder inside the dipstick is varied continuously by introduction of a contact gas. The low-temperature bias tee decouples the high frequency path from the DC bias path (Fig. 5.1). The high-frequency signal is led out of the dipstick by stainless-steel rigid coaxial cables, while DC bias is provided via a pair of twisted wires. The samples are biased by a battery-powered low-noise DC source. The signal is amplified at room temperature by several amplifiers with the total gain of 70 dB and then send to a pulse counter with a 300 MHz physical bandwidth.

The optical fiber feeds the light from the monochromator into the cryogenic part and is mounted on a movable stage above the sample surface. The distance between the end of the fiber and the surface of the detector is about 4 mm. This ensures that the size of the light spot on the detector is much larger than the wired area (4-7 µm). In this case, the distribution of the light intensity is homogenous across the wired area.

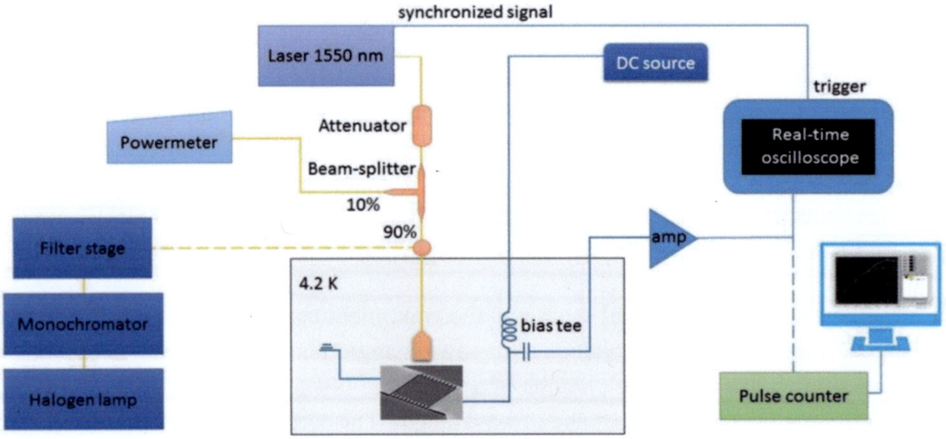

Figure 5.1: Scheme of experimental setup for the measurement of detection efficiency of SNSPD in a spectral range from 400 – 2200 nm and timing jitter of detector. The optical fiber (yellow) feeds the light from the monochromator (or laser) into the cryogenic part and is mounted on a movable stage above the sample surface. The RF and DC paths are de-coupled at the sample stage and led out separately. The voltage pulses of the detector are amplified at room temperature before they are recorded by a pulse counter. The power of incoming light from laser is controlled by powermeter and could be adjusted via variable attenuator. The electrical synchronized signal from laser is delivered to real-time oscilloscope for triggering of voltage pulses from detector.

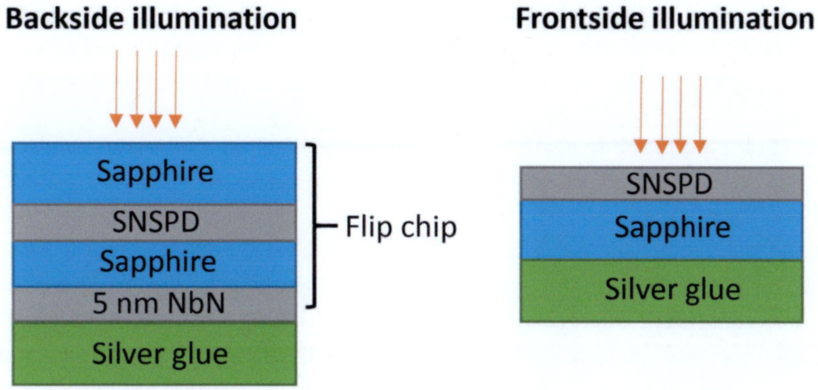

Figure 5.2: Optical assemblies which were used for measurements of the detection efficiency. The flip-chip package technology is used to mount single-spiral SNSPD in the sample holder (left scheme). The single-layer SNSPDs in form of meander and double spiral were placed in standard configuration (right scheme).

For measurements of the optical response, detectors with the meander and the double-spiral layouts are illuminated from the front side (right sketch in Fig. 5.2). Because the nanowire in the single spiral layout is covered by an AlN isolating layer and a Nb-top electrode, which reflects light, the detectors with this layout are illuminated from the rear side, i.e., through the non-polished sapphire substrate. To eliminate the difference in the film absorbance for different illumination scenario and hence to avoid normalizing the measured detection efficiency for the absorbance, the special flip-chip assembly (shown in the left sketch in Fig. 5.2) is designed. Taking the absorbance of 5 nm NbN film on one-side polished sapphire illuminated from front and back sides (presented in section 3.1.1) into account, the absorbance of optical assemblies of SNSPD has been estimated. It is straightforward to show that for a thin film the film absorbance in two assemblies is the same with an accuracy better than 15%.

5.2 Influence of geometry on performance of SNSPD

5.2.1 Superconducting properties

In order to compare the superconducting properties of fabricated detectors with different layouts, the measurement of temperature dependence of resistance of all specimens have been done from 4.2 up to 300 K by a quasi-four-probe technique. Fig. 5.3a represents examples of *R*(*T*)-curves of SNSPDs with different layouts and single strip. The single strip which was patterned in one process with detectors is considered as a reference structure.

After patterning the film into a nanowire (technology of single- and multilayer SNSPDs is described in section 3.3), the critical temperature of the nanowire becomes smaller than the T_C = 13 K of the non-patterned film. The key parameters of detectors are compiled in Table 5.1. Among specimens with different layouts the single bridge has the highest critical temperature 12.8 K (green curve in Fig. 5.3a). The slight variation of T_C was found for meanders, double spirals and single spirals in the range between 12 and 12.4 K. The difference in T_C of detectors and single strip can be associated with defects which are presented in the film. Since the length of the strip (2 μm) is more than one order magnitude shorter in the comparison to the length of detectors (80 μm and 200 μm for meander/double spiral and single spiral respectively), the probability of defects in the strip is much lower. Indeed, it was found that the critical temperature of strips with variation of the length and same nominal width (100 nm) reduces in longer strips. In Fig. 5.4, the critical temperature of strips decreases from 12.9 down to 11.8 K in the range of the width from 200 nm up to 40 μm.

The current-voltage (CV) characteristics of detectors were measured in the current-bias mode at 4.2 K (Fig. 5.3b). Despite of structures have nominally identical width of naowire, a variation of I_C values is observed. Similar to dependence of T_C, the highest value of the critical current (46.6 μA) was measured in single strip (green curve in Fig. 5.3b). Among SNSPDs, single-spiral detectors have highest $I_C \approx 39$ μA (black curve in Fig. 5.3b). Double-spiral and meander detectors show lower critical current about 34 μA and 28 μA, accordingly.

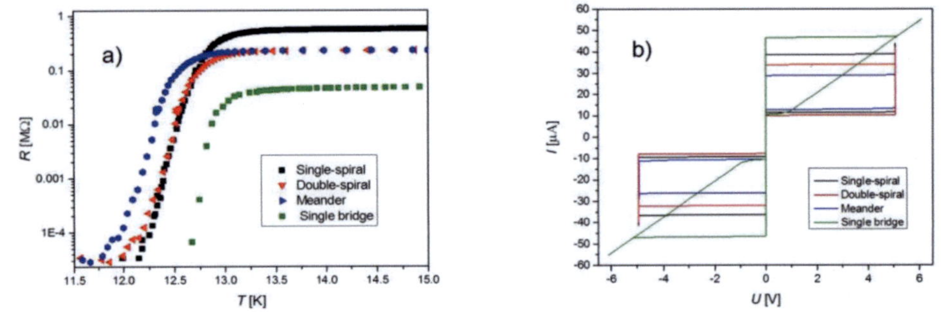

Figure 5.3: a) Temperature dependence of resistance of SNSPDs with different layouts (indicated in a legend). The single strip is a bend-free reference structure. b) Current-voltage characteristics of several types of SNSPDs and single strip (indicated in a legend) measured in current-bias mode at 4.2 K.

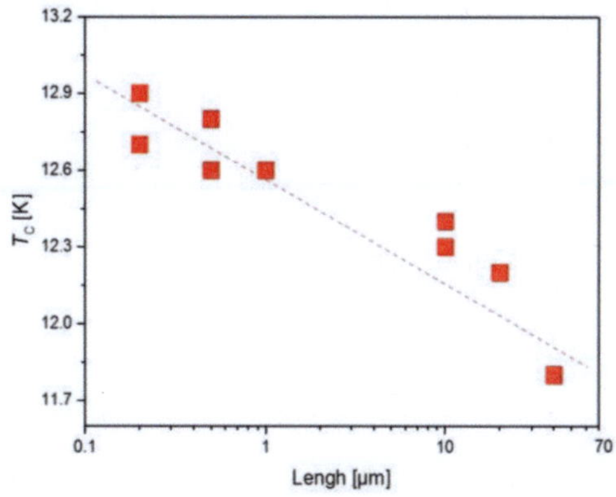

Figure 5.4: Critical temperature as a function of length of strip. The nominal width of strips is 100 nm. The dashed red line guides the eyes.

To compare experimental critical currents with the theoretical limit, the de-pairing critical current $I_C{}^d$ was computed in the framework of the standard Ginzburg-Landau (GL) approach with the Ginzburg-Landau temperature dependence and the correction for the extreme dirty limit (Eq. 4.17):

$$I_C^d(T) = \frac{4\sqrt{\pi}\left(e^{\gamma}\right)^2}{21\varsigma(3)\sqrt{3}} \frac{\beta_0^2 \left(k_B T_c\right)^{\frac{3}{2}}}{eR_s\sqrt{D\text{h}}} w \left[1-\left(\frac{T}{T_c}\right)\right]^{\frac{3}{2}} \times K\left(\frac{T}{T_c}\right), \tag{5.1}$$

where $\beta_0 = 2.05$ is the ratio of the energy gap at zero temperature to $k_B T_C$. The maximum ratio of the measured I_C to the $I_C{}^d$ was 0.55, 0.48 and 0.4 for the single spiral, double spiral and meander layouts, respectively.

The critical current density $j_C = \frac{I_C}{wd}$ of samples with different layouts is displayed in Fig. 5.5. The highest value of 8.4 MA/cm^2 was obtained with a single strip. The critical current density of the meander was the lowest among these samples. The j_C of meander is 38 % lower than the j_C of the straight bridge with the same width. The decrease of the critical current density in spiral and meander nanowires is caused by a lower T_C (Fig. 5.3a) and by current crowding in bends.

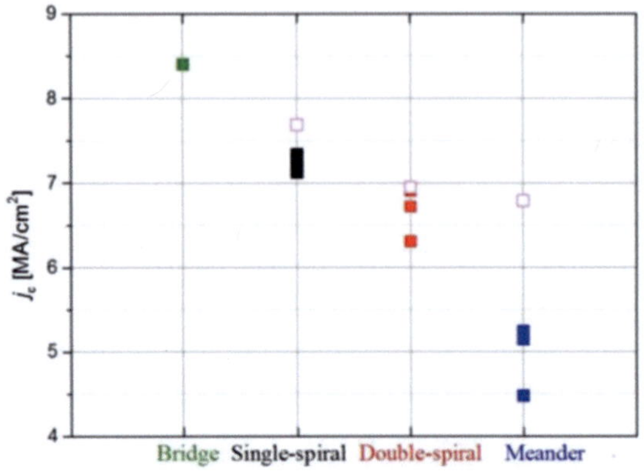

Figure 5.5: Critical-current densities j_C(4.2 K) of single bridge and nanowires with single- and double-spirals and meander-type layouts. The opened pink dots show the values scaled to the same relative temperature $T/T_C = 0.33$ for each layout.

To account for different critical temperatures, the critical current densities in Fig. 5.5 was scaled to the same relative temperature $T/T_C = 0.33$ (smallest value corresponding single bridge) by means of the Bardeen-temperature dependence of the critical current [94] (opened pink points in Fig. 5.5). Thus, the decrease of the critical-current density in meander due to its geometry is about half of the difference in the critical-current densities between the single strip and the meander at 4.2 K. The spirals demonstrate higher j_C in comparison to the meander. Despite the absence of sharp bends in single spirals, their averaged $j_C = 7.2$ MA/cm^2 is below the value which was measured for a straight bridge. At the same time, j_C of the double spiral is slightly lower than the value of the single spiral. It is commonly accepted [60], [62] that non-uniformities, constrictions and bends restrict the experimental critical current in nanowires. For the layouts studied here, there are turns in the meander and the sharp turn in the center of the double spiral where current crowding occurs. The straight nanowire, which has no bends, demonstrates the highest density of the critical current. Although the single-spiral layout is free from bends, j_C remains smaller than in the straight nanowire. One of possible reason can be the inhomogeneity of nanowires which was studied systematically on nanowires with different lengths [95]. The observed trend of reduction of j_C with increase of the length of nanowire has been conditioned by inhomogeneous nanowires at a length scale shorter or equal to 100 nm. Furthermore, the winding radius of the nanowire in the center of the single spiral could be small enough to cause weak current-crowding effects.

The equilibrium between power of dissipated Joule heating in nanowire and cooling efficiency through nanowire-substrate interface determines the retrapping current, j_r (Table 5.1). The slight variation of j_r is due to difference in width of nanowires and inaccuracy of corresponding measurement of the width via SEM.

Table 5.1: Parameters of detectors and reference structure made of superconducting 5 nm thick NbN film: width of nanowire W, critical temperature T_C, critical current I_C, retrapping current I_C, critical current density j_C, density of retrapping current j_r.

Type	Name	W, nm	T_C, K	I_C, µA	I_r, µA	j_C(4.2 K), MA/cm^2	j_r(4.2 K), MA/cm^2
DETECTORS	Single spiral	111	12.4	39.5	12.7	7.1	2.3
		108	12.2	39.1	12.6	7.2	2.3
		105	12.2	38.5	11.4	7.3	2.1
	Double spiral	102	12.15	34.2	10.7	6.7	2.1
		100	12.1	33.9	9.99	6.9	2
		107	12.3	34	12.6	6.3	2.4
	Meander	110	12.1	28.8	11.8	5.2	2.2
		111	12.15	28.6	12.6	5.1	2.3
		101	12	22.6	9.1	4.5	1.8
Reference structure		110	12.8	46.6	10.5	8.4	2

5.2.2 Spectral detection efficiency

The spectral detection efficiency, *DE*, for all types of detectors (see table 5.1) can be measured from 400 up to 2200 nm wavelength for different bias currents at the experimental conditions described in section 5.1. For the particular wavelength, *DE* is defined as the ratio of the rate of photon counts to the rate of photon arrival to the illuminated side of the specimen.

The spectral dependences of the *DE* at different relative bias currents are shown in Fig. 5.6.

Each $DE(\lambda)$ curve demonstrates a clear roll off which begins around a particular wavelength. The cut-off wavelength λ_C is defined as the wavelength corresponding to the intersect of two straight lines which extrapolate the plateau at small wavelengths and the decaying portion of the $DE(\lambda)$ curve (Fig. 5.6a). At the same relative bias current $I_b = 0.95 I_C$ (Fig. 5.7), the single spiral demonstrates λ_C in the range of 900 nm, while the double spiral and the meander have smaller λ_C ($\approx$ 600 nm).

On the plateau, the detection efficiency of single spirals reaches 29.6%. Scaling this value to the filling factor of 50% of other layouts, it arrives at numbers close to the plateau efficiency 23.4 % of the double spiral. In contrast, the meander has the plateau efficiency of only 15.7%.

When the bias current increases, *DE* grows much stronger at wavelengths beyond λ_C than at the plateau. Correspondingly, λ_C increases with the current.

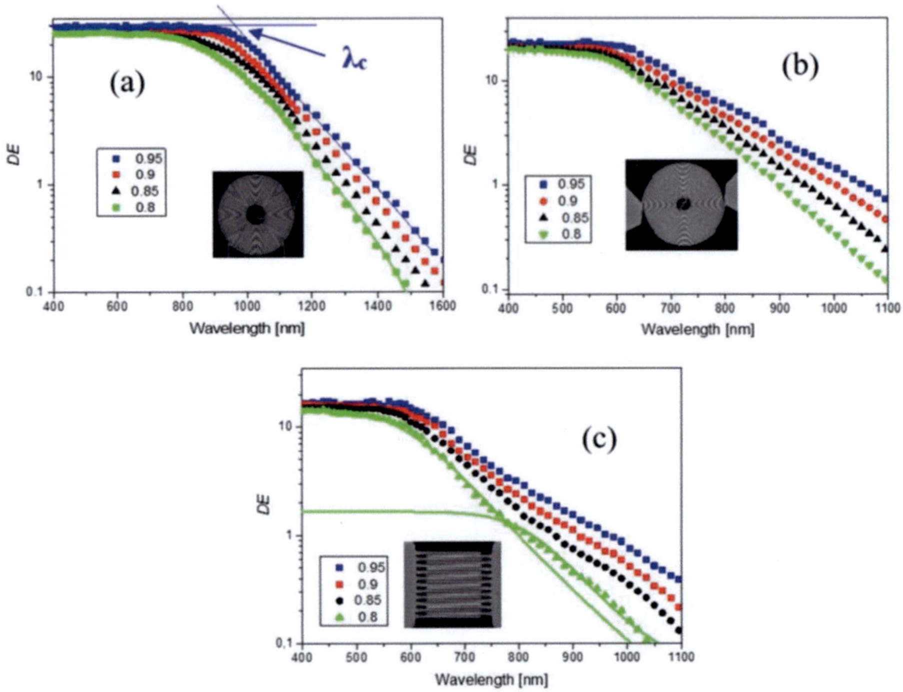

Figure 5.6: Spectral detection efficiencies of detectors with single-spiral (a), double-spiral (b) and meander (c) layouts for different relative bias currents (shown in the legends). Solid green curves in panel (c) show best fits which were made with Eq. (5.2).

The almost current-independent detection efficiency on plateaus for all three layouts (Fig. 5.6) represents the maximum efficiency when the whole nanowire together with turns contribute to the photon-count rate. The roll-off begins when the central portion of the nanowire becomes inactive in detecting photons [24]. At even larger wavelengths, mostly bends and turns contribute to the photon-count rate [SCL+15]. The transition from the plateau to the decaying part of the *DE* spectrum is formally described by:

$$DE = \left(1 + \left(\frac{\lambda}{\lambda_C}\right)^p\right)^{-1}, \tag{5.2}$$

where λ_C is the cut-off wavelength and the exponent p describes the power-law decrease of the efficiency in the near infrared range. The green solid curve in Fig. 5.6a shows the best fit of the spectral detection efficiency with the Eq. (5.2) which was obtained with λ_C= 749 nm. This value coincides with λ_C which is obtained by approach shown on Fig. 5.6a.

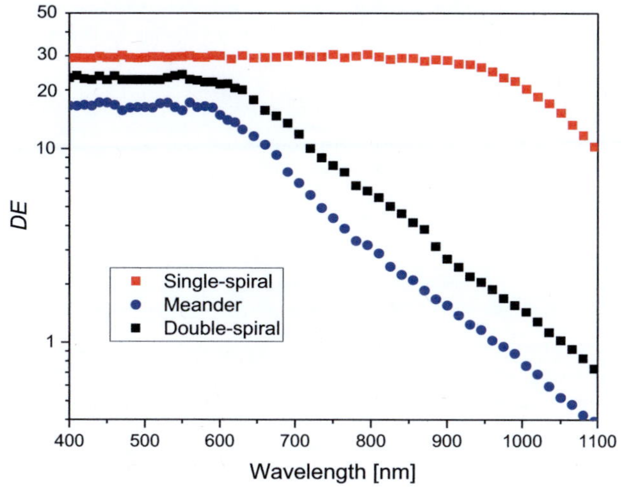

Figure 5.7: Spectral detection efficiencies of detectors with single-spiral, double-spiral and meander (indicated in legend) layouts for $0.95I_C$ relative bias currents.

The spectral detection efficiency of the meander-type specimen shows a second pronounce roll off for all relative bias currents (Fig. 5c). Beyond the cut-off wavelength, *DE* decreases monotonically until at some point the *DE* jumps up (intersection of two green solid curves in Fig. 5.6c). This roll-off is associated with detection of photons by turns while the straight portions are less active than bends at wavelengths larger than at λ = 760 nm. Two solid curves depict fitting functions $DE(\lambda_1)$ and $DE(\lambda_2)$ which were obtained with Eq. 5.2. They represent the detection efficiency of straight and bended parts of meander, respectively. The detection efficiency of the meander-type specimen can be presented as:

$$DE(\lambda) = DE(\lambda_1) + DE(\lambda_2) \tag{5.3}$$

From the ratio of fitted detection efficiencies at plateaus of $DE(\lambda_1)$ and $DE(\lambda_2)$, areal ratio of bends and straight parts of the meander was estimated. The equivalent number of squares in bended area amounts to 11% of the number of bends. This is very close to the geometric estimation of this ratio.

For all layouts the cut-off wavelength increases with bias current (Fig. 5.6). In the framework of the model of diffusive hot-spot (section 2.2.1), the dependence is linear (Eq. 2.5), while the steady-state hard-core model (Eq. 2.3) predicts a different dependence of the cut-off wavelength on the bias current. For all three layouts, the relative bias current is plotted in Fig. 5.8 as function of the photon energy corresponding to the cut-off wavelength measured at this particular current along with best fits by three different models.

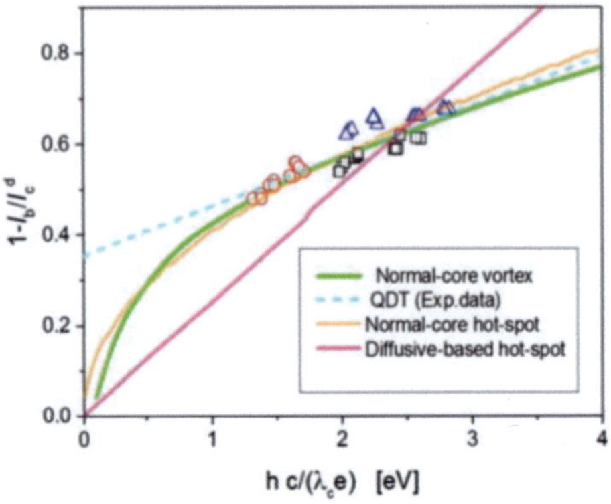

Figure 5.8: Inverted relative current $1\text{-}I_b/I_c^d$ versus energy of photons at the cut-off wavelength measured at this current. Symbols show experimental data for all three layouts. Lines show the best fits obtained with different theoretical models. The models are specified in the legend.

For all fits the effectiveness ζ is used as the only fitting parameter and the same superconducting and material parameters. The burgundy straight solid line represents the fit from the diffusion hot-spot model with $\zeta = 0.6$, the black dash dotted line – from the hard-core model with $\zeta = 0.15$ and the violet curved solid line is the best fit from the numerical model invoking Ginsburg-Landau approach to the evolution of the order parameter [96] with $\zeta = 0.09$. The dimensionless wire width of nanowires $w/\xi = 19$ is appropriate for this fit. For comparison, the straight dashed line shows the phenomenological linear approximation from Ref. [53]. As it is clearly seen, the GL approach provides the best match with the experimental data simultaneously for all specimens. The fact that all experimental points fall on the same theoretical curve evidences the key role of the experimental critical current in the position of the spectral cut-off. This fact also confirms the supposition that the cut-off is caused by the position dependent probability of photons to be detected in a straight line and that it is almost unaffected by the bends in the layout.

5.2.3 Dark count rate and timing jitter

Dark counts

The dark-count rate, *DCR*, is determined by blocking the optical path at the fiber input. While *DCR* of the single-spiral SNSPD stayed below 10 cps for all relative currents, the double spiral and meander showed an increasing *DCR* up to 10^3 cps at the relative bias current $0.95 I_C$ (plotted as function of I_b/I_C^d in Fig. 5.9). Dark counts weren't observed at relative bias current lower than $0.75 I_C$.

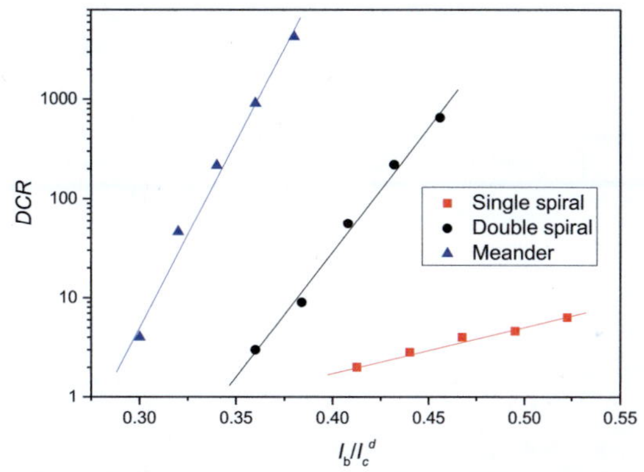

Figure 5.9: Dependence of the dark-count rate (*DCR*) on the relative bias current for three layouts which are indicated in the legend. The solid lines are fits by formalism of Bulaevskii [52]

The different rates of dark counts in detectors support the envisaged role of turns [SCL+15]. Indeed, the meander contains turns with the smallest radius and has, correspondingly, the highest rate of dark counts. The smallest rate was observed in single spirals where the winding of the nanowire has the largest bending radius. The intermediate rate of dark counts in double spirals is easily attributed to sharp turn of the current in the center of the layout where the nested spirals are connected to each other. The formalism of Bulaevskii et al. [52] is used to fit the dependences of *DCR* on bias current (Eqs. 51, 52 in Ref. [52]). The fits are shown with straight lines in Fig. 5.9.

Timing jitter

Timing jitter, which is the probability density of the random time delay between photon absorption in current-carrying superconducting nanowire and appearance of the normal domain, reveals two different underlying physical scenarios in the deterministic and probabilistic regimes [SSH+17]. The measure of the timing jitter is the width of the statistical distribution in the arrival times of voltage transients with respect to the corresponding photon absorption times.

To measure timing jitter, a 1560 nm femtosecond-pulsed laser (C-Fiber, Menlo Systems [97]) was adopted as the single-photon source in the experimental setup (Fig. 5.1). The 32-GHz real-time oscilloscope builds a statistical distribution of the arrival times of photon counting pulses. The distribution typically has a Gaussian profile as presented in Fig. 5.10 for double-spiral detector biased at different currents. The timing jitter is defined as the full width at half maximum (FWHM) of the distribution. The jitter increases when the bias current decreases. This is also illustrated in Fig. 5.11a. At relative bias current $0.95 I_C$, the system jitter (FWHM) is found as σ_{system} 49 ps, 38.6 ps and 42.7 ps for single spiral, double spiral and meander, respectively.

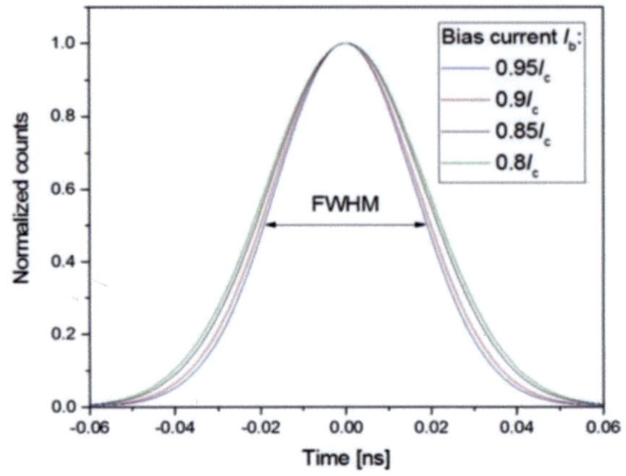

Figure 5.10: Distribution of the signal arrival times for the specimen with the double-spiral layout biased at currents indicated in legend. The minimum jitter (FWHM of the distribution) amounts at 38.6 ps for relative bias current $0.95I_C$.

The jitter is normalized for the actual length of the nanowire in the particular design to account for different length of detectors. The obtained ratio of jitter per unit length is plotted as a function of relative current ratio I_b/I_C^d in Fig. 5.11b. The single spiral has the smallest jitter per unit length and the weakest dependence of the jitter on the bias current. The jitter value and the steepness of this dependence increases for the double spiral and further for the meander.

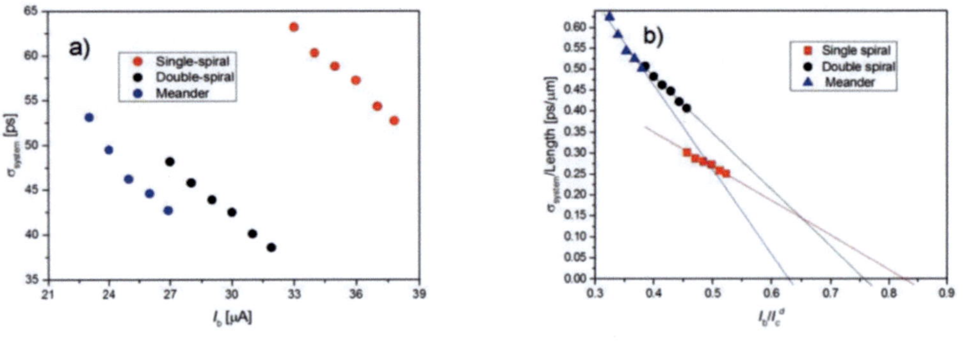

Figure 5.11: a) System jitter σ_{system} versus absolute bias current for single-spiral, double-spiral and meander layouts. b) Timing jitters per unit length of different nanowires as function of the relative current for layouts indicated in legend. The solid lines guide the eyes

The system jitter can be presented as the mid square of several components:

$$\sigma_{system} = \sqrt{\sigma_{SNSPD}^2 + \sigma_{noise}^2}, \tag{5.4}$$

where σ_{SNSPD}, σ_{noise} are the jitters from the detector and readout circuit of measurement setup, respectively.

The jitter produced by electrical circuit, σ_{noise}, can be estimated using single time-trace (Fig. 5.12) as:

$$\sigma_{noise} \approx \tau_{rise}\left(\frac{\sigma_N}{0.8A_{max}}\right), \tag{5.5}$$

where τ_{rise} is a pulse rise time taken between 10 and 90 % of pulse amplitude; σ_N is a root mean square noise amplitude; A_{max} is a maximal pulse amplitude. Taking σ_{noise} into account, the σ_{SNSPD} was obtained using Eq. (5.4) as 32 ps for single spiral, 18.4 ps for double spiral and 22.5 ps for meander at $I_b = 0.95I_C$.

The observation of the different timing jitter per unit length in specimens with different layouts supports the concept [98] that the nanowire can be thought of as a transmission line with its specific impedance l/v where l is the length of the nanowire and v is the group velocity. Hence, the values shown in Fig. 5.11b are the reciprocal group velocity. Apart from the dispersion, the group velocity equals the phase velocity and depends on inductance of the line per unit length L. For the superconducting transmission line, the inductance is primarily given by the kinetic inductance of the nanowire [DKG+17].

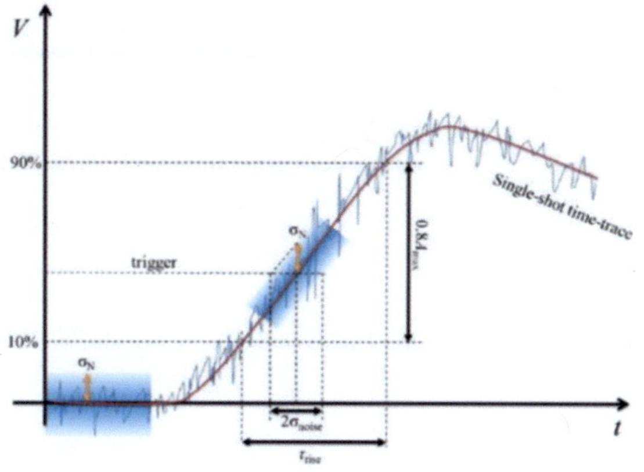

Figure 5.12: The single-short time-trace of SNSPD. The trigger level is set in 50 % of maximum pulse amplitude, A_{max}. The pulse rise is defined by time taken in range 10-90% of A_{max}.

5.3 Summary

In this chapter, SNSPDs shaped as either meander or spirals are systematically characterized and compared. The spirals had different layouts, a double-spiral layout with an S-turn in the middle and a single-spiral layout without such turn. Nanowires were prepared from films of niobium nitride with a thickness of 5 nm as described in Chapter 3. For detectors of each layout, the spectra of the single-photon response were measured in the wavelength range from 400 nm to 1600 nm and defined the cut-off wavelength (λ_C) beyond which the response rolls off. The largest and the smallest λ_C were found for the single-spiral layout and for the meander, respectively. For all three layouts the relationship between λ_C and the relative bias current falls onto a universal curve which has been predicted in the framework of the modified hot-spot model [96]. It was shown that avoiding bends in a current-carrying superconducting nanowire enhances the probability for low energy photons to be detected and that this enhancement is entirely due to the increase in the experimentally achievable critical current. For single-spiral detectors, the efficiency of photon detection at wavelengths smaller than λ_C reaches the expected absorbance of the spiral structure and the timing jitter per unit length of the nanowire has the smallest value. In terms of the detection efficiency and spectral cut-off the most important scaling factor is the critical current for the particular layout while the bends play a minor role. Contrary, the rate of dark counts as well as the timing jitter are mostly restricted by the quality and number of bends in the layout. The value of timing jitter is directly related to the length of the nanowire, its kinetic inductance and the amount of bends.

6 Optical response of single-spiral SNSPD in the magnetic field

The vortex movement is of particular importance in current-biased thin-film superconducting nanowires. It was assumed that vortices can contribute to the formation of dark count events [99]. Furthermore, the detection of near-infrared photons ($\lambda > \lambda_C$) in superconducting meander detectors can possibly be explained by penetration and movement of vortices in nanowire [100]. In the previous chapter, it was shown that the spectral detection efficiency decreases exponentially for photons with wavelengths $\lambda > \lambda_C$. If the photon energy is not sufficient to cause a normal region over the entire strip width, single-photon detection is related with the movement of vortices. The superconducting energy gap is locally reduced by the absorption of low-energy photon. This results in an increased probability of overcoming the barrier through the vortex at this location and its subsequent crossing of the strip. If the vortex crosses the strip, it releases sufficient energy to locally destroy the superconducting state and to generate a measurable voltage pulse. The vortex-assisted model predicts different magnetic field dependencies of *PCR* and *DCR*. Furthermore, the magnetic field should not influence on the formation of the hotspot by photons, but initiates a subsequent single-vortex crossing, which provides the rest of the energy needed to create the normal-state belt. From another point of view, an applying of external magnetic field to bended superconducting structures enhances the critical current and corresponding the cut-off wavelength of the detector.

In order to investigate the influence of the magnetic field on the *PCR* and *DCR* in SNSPDs and to investigate the role of vortices in the photon detection process and dark count events, the experimental setup with a superconducting coil is realized and measurements of the *PCR* and *DCR* in a magnetic field is performed on detectors appropriate for applying external magnetic field. In this experiment, SNSPDs in forms of square and circular spirals with turns of a single symmetry are used. Additionally, the $\lambda_C(B)$-dependence is studied by measurement of spectral detection efficiency in the magnetic field.

6.1 Cryogenic setup with superconducting coil

This section provides a general description of the experimental setup with a superconducting coil. The experimental setup for measuring the *PCR*s and *DCR*s of SNSPDs as a function of the magnetic field is shown in Figure 6.1. The measurement setup includes a dipstick with a superconducting coil and several readout components for operation at room temperature.

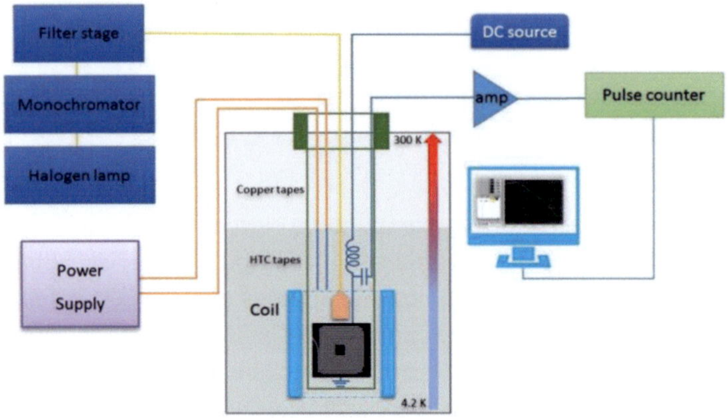

Figure 6.1: Simplified sketch of the experimental setup for measuring the *PCR* and *DCR* of SNSPDs as a function of the magnetic field.

The principle of cryogenic measurement setup is similar to the setup described in previous chapter. However, some parts of the dipstick with superconducting coil are more complicated than it was used for spectral detection efficiency. The dipstick with magnet, which is immersed in a Dewar with liquid helium consists of an insert with a superconducting coil and a vacuum sample rod on which the SNSPDs are installed (Fig. 6.2a). The pressure in the vacuum sample rod is adjusted by controller. In order to lead the current from the magnet power supply (4G) to the superconducting coil with minimum energy dissipation, the electrical path is divided into two parts. The copper tapes integrated in the dipstick (Fig. 6.3b) lead the current from room down to stage of the coupling with HTC tapes covered by thin copper stabilizer. The usage HTS tapes allows to reduce a total heating due to high current though bias lines. The superconducting tape has a critical current of 84 A at 77 K which is high enough for operation of the superconducting coil.

However, the ambient temperature should be lower than T_C of high-superconducting tape.

The temperature sensor is mounted on the stage of coupling for control of the ambient temperature. The bias tee is embedded in a compact box for operation at low temperature (Fig. 6.2c). Taking the required frequency range of SNSPD signals (between several MHz and several GHz) into account, the bias tee was designed using a capacitance of 12 nF and a coil inductance of 840 nH.

The superconducting coil is specifically designed in accordance with requirements of dimensions of components in experimental setup. First of all, the outer diameter of the coil is limited by the diameter of the neck of the standard 60/100-liter He^4 transport Dewar.

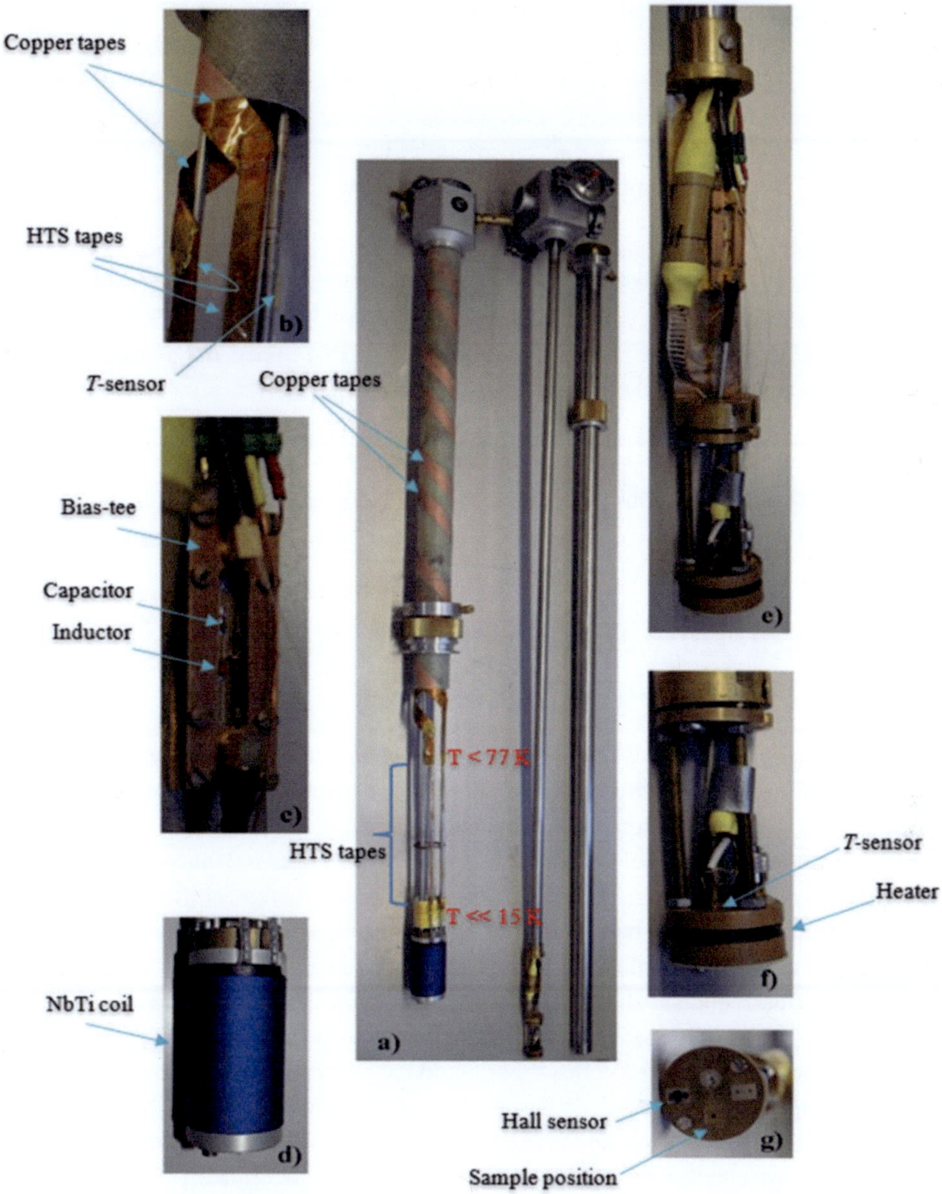

Figure 6.2: a) Variable temperature insert with a superconducting coil and vacuum sample rod. The vacuum sample rod is assembled in the dipstick with coil during operation. b) The stage of coupling of the copper and high-superconducting tapes. c) Embedded bias-tee d) Superconducting NbTi coil with central field up to 2 T at 4.2 K. e) Carrier board for electrical and optical components. f) Sample holder with embedded temperature sensor on the back side, heater and SMA fiber connector for back side illumination. g) View of top side of the holder with Hall effect sensor for control magnetic field.

This restricts maximum central field of superconducting coil. The custom-designed coil is designed for 2 T at 4.2 K operation. The operating current applied to the coil at the maximum of field is 49.3 A. In this case, field-to-current ratio is about 0.041T/A. In order to control the magnetic field along the axis of the superconducting coil and measure the actual field near the sample a calibrated Hall sensor with ultra-high sensitivity and wide measuring range (1 mT - 2T) was mounted on the sample holder.

The sample holder was designed to fit the inner diameter of the superconducting coil which is 31 mm. The PCB for the sample holder was done with a high-resolution photo-lithographic mask and subsequent wet-etching process of printed circuit board (RogersTMM10i substrate with 35 μm copper layer on both sides). The line impedance is matched to the readout components and is therefore 50 Ω. The heater is easily made by wiring of constantan wires around the sample holder (Fig. 6.2f). The field-independent temperature sensor (Lake Shore CX-1010-SD-HT-1.4L) is embedded in a sample holder. The electrical signal from the sensor and other electrical components (heater, Hall sensor) is guided out via twisted pairs (Fig. 6.2e).

6.2 Magnetic-field dependence of the performance of SNSPD

Current crowding in the bends of superconducting nanostructures not only restricts measurable critical current in such structures, but also redistributes local probabilities for the appearance of dark and photon counts. Statistical fluctuations in the form of dark counts restrict the minimum detectable photon flux. At the inner edges of bends and turns, where the supercurrent rounds a sharp corner, local current density increases, which causes a local reduction of the free energy barrier for nucleation of magnetic vortices. Among different topological fluctuations, hopping of vortices across the strip is commonly considered as a mechanism of dark counts. Hence, bends and turns with sharp corners are places from which dark counts most probably come. Photon counts in narrow strips are related to either current- or fluctuation-assisted vortex crossing. In the first deterministic scenario, a photon creates a hotspot in the strip, which forces the current density to redistribute around the absorption site. A vortex nucleates at any point where, after current redistribution, the velocity of the superconducting condensate locally achieves its critical value. Discovering the local nature of count events has made it possible to bridge these two scenarios in the framework of the deterministic model. However, strips in the common meander form prevent one from figuring out where count events occur. Differentiating contributions from straight portions and bends by applying a magnetic field runs into the problem that the meander has bends with a different symmetry with respect to the directions of the current flow and magnetic field. Chirality, i.e. the dual symmetry of turns in the meander, masks the expected asymmetry in magnetic fields for count events that occur in bends. Using single bends and bridges helps to solve the problem but has its own complications, such as resonance effects in the absorption probability for particular wavelengths and current crowding imposed by closely spaced contacts. Furthermore, optical coupling to small structures is deteriorated.

In this section, the optical response of specially designed square-shaped spirals is studied. They contain bends with the only one symmetry with respect to current and magnetic field directions and have an optical coupling efficiency comparable to the meandering strips. As a reference, circular spirals without bends are used. The manufacturing process of spiral structures and their characterization are described in previous chapters.

6.2.1 Critical current

The depairing current is the critical parameter of the superconducting state, which provide scales for measured critical currents and applied fields. The depairing current in straight portions of the square spiral was computed in the framework of the standard Ginzburg-Landau (GL) approach with the Bardeen temperature dependence and correction for the extreme dirty limit (Eq. 5.1) with parameter of given film R_{sq} = 300 Ω at 20 K, D = 0.5 cm^2/sec is the typical diffusivity of normal electrons in NbN films [16]. The width of nanowire was measured via SEM inspection.

For the square spiral with W = 110 nm and T_C = 11.8 K, the depairing critical current of 129 μA was estimated at T = 4.2 K. The experimental critical current of the square spiral is measured in the range of 36 μA.

It is expected that, in a square spiral, current crowding at the inner corners of bends will reduce the measured critical current of the whole spiral with respect to the critical current of its straight parts. An external magnetic field induces screening current in bends. Depending on the field direction, the screening current may decrease or increase the local current density at the inner corners of bends [65]. Figure 6.3 shows a combination of field and current directions, which results in an increase of the local current density at the inner corner of a bend.

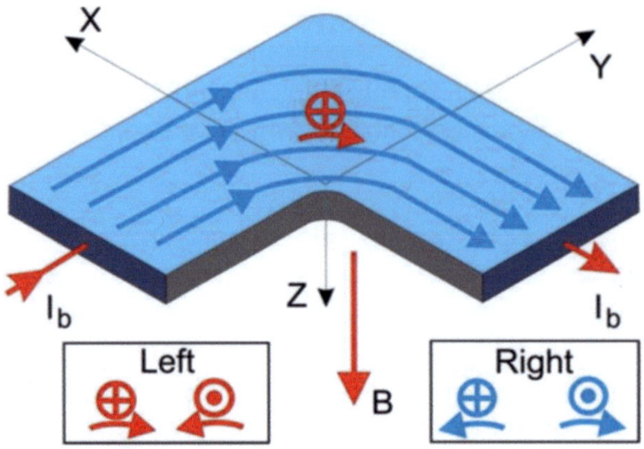

Figure 6.3: Bend schematics and positive directions of the external magnetic field (B) and bias current (I_b). These directions obey left field-current symmetry for which an increase in the magnetic field causes the increase in the superconducting current density at the inner corner of the bend. Pictograms in the left box denote two possible configurations that produce this effect. Pictograms in the right box denote configurations having right symmetry and, consequently, opposite effect on the current density at the inner corner. The pictogram in the bend corresponds to the positive directions of the field and current shown in the figure. The inner corner has coordinates (0; 0; 0) in the system shown here.

The sign of the effect remains unchanged when both field and current directions are changed to the opposite. The direction combinations, which have such an effect on the current density, is called the combination with the left field-current symmetry. Two pictograms in the left box depict two combinations with the left symmetry. The crosses or points in the circles denote two opposite directions of the magnetic field and the arrows denote the directions of the bias current in the bend. The other two combinations will be called combinations with the right field-current symmetry. The corresponding two pictograms are shown in the right box. Pictograms will be using through the chapter to relate experimental data on plots to specific combinations of field and current directions. In order to visualize the expected effect of the magnetic field on the current density, the local density of superconducting current in a bend was computed in the framework of the GL formalism [101] without a magnetic field and at a field of $B = \pm 0.005 B_{c2} = 66$ mT [102]. The results are shown in Fig. 6.4 as two-dimensional contour plots. The critical current is achieved when either the vortex barrier at the inner corner disappears [52] or the local current density at the inner corner equals the depairing current density [103]. In both cases, one expects the critical current to decrease with an increasing field for the left field-current symmetry and to increase for the right field-current symmetry. The critical current was measured for all possible combinations of field and current directions as a function of the magnetic field. The results are shown in Fig. 6.5 for (a) the square spiral and (b) the circular spiral. The square spiral demonstrates dependences expected for a single bend.

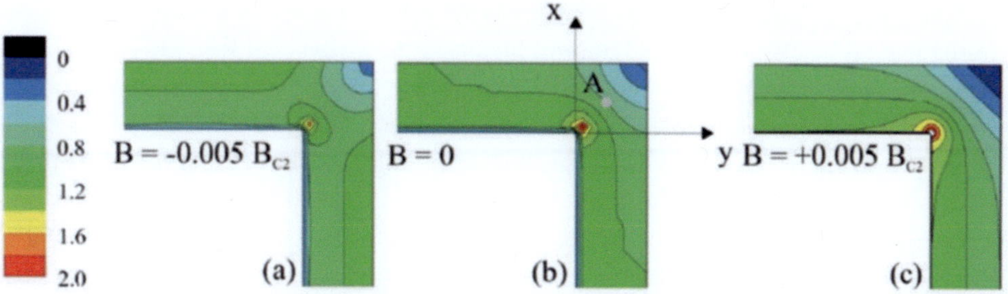

Figure 6.4: Density of the superconducting current in the bend and in the adjacent straight parts of a strip (b) without magnetic field and for the field $0.005B_{c2}$ with (c) the positive and (a) negative directions. The current density is normalized to its density far from the bend at $B = 0$. The current distribution was computed for the strip width of 20ξ where $\xi = 5$ nm is the coherence length for NbN film. A gray circle labeled with the letter A in the common bisector of the bend corners (panel (b)) marks the position where the local current density does not change with the magnetic field.

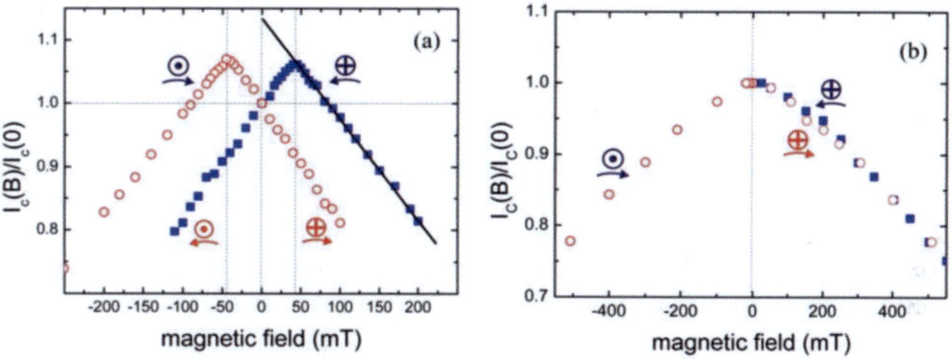

Figure 6.5: a) Relative critical current of the square spiral in magnetic field for positive (open symbols) and negative (closed symbols) current directions. Pictograms depict combination of the field and current directions for each section of the plot. Solid straight line extrapolates to zero field the linear decrease of the critical current with the magnetic field in the right symmetry. Vertical dashed lines show zero field and positions of the maxima on the field axis. b) Relative critical current of the circular spiral for different current directions. The same convention is used to mark symmetries and current directions.

For combinations with the right symmetry, the critical current increases with the field, reaches a maximum at $B_{max} = 44$ mT, and further decreases. For combinations with the left symmetry, the critical current linearly decreases with the magnetic field. This effect was already reported for separate bends [66]. Simultaneous change of current and field directions mirrors the $I_C(B)$ curves with respect to the $B = 0$ line.

Circular spirals do not show any asymmetry of the critical current in a magnetic field. Assuming that all bends in square spirals are identical, the analysis of Ref. [16] was applied to find, via linear extrapolation of the field dependence for the right symmetry, the critical current in the straight parts $I_{C0} = 42\mu$A and the reduction factor $R = I_{Cm}/I_{C0} = 0.86$ due to current crowding. Here, I_{Cm} is the maximum experimental critical current in magnetic fields. The critical current in straight parts of the spiral is less than the computed depairing current. The difference is within the range found for nanowires with similar stoichiometry [16]. The self-field that is produced by the critical current in the middle part of spirals is less than 0.1 mT, which is from two to three times larger than the local earth magnetic field and almost two orders of magnitude less than typical B_{max} values.

The critical current in square spirals, i.e. the current at which the superconducting state becomes nonstable, was found from the numerical solution of the GL equations [101] in the geometry shown in the inset in Fig. 6.6. The bends (B) where rounding is neglected, and straight segments of the strip with edge defects (A) are separately considered. The results are presented in Fig. 6.6 separately for the bend and for the straight segment. It was found that the maximum of the critical current in the bend (closed symbols in Fig. 6.6) should occur for the right symmetry at $B \approx 0.02B_{c2} \approx 260$ mT. This is almost twice as large as the value obtained with the London model (Eq. (17) in Ref. [65] for a sharp 90° bend. Taking into account nominal rounding of inner corners in bends of structures $r/W = 0.65$ and assuming that all bends are identical, it expects in the framework of the London model that a reduction factor of $R = 0.75$ (Fig. 14 in Ref. [60] for the critical current in bends and the maximum in the experimental critical current at $B_{max} = 65$ mT (Eq. (17) in Ref. [65]). The experimental values $R = 0.86$ and $B_{max} = 44$ mT (Fig. 6.5) are reasonably close to the predictions of the London model. Moreover, for the experimental reduction factor $R = 0.86$, the London model predicts $B_{max} = 38$ mT, which even better corresponds to the obtained experimental value $B_{max} = 44$ mT. The remaining difference between the experimental reduction factor and the predicted factor, which is expected for a nominal rounding in structures, attributes to geometrical nonuniformities of the strip edges. Such nonuniformities typically appear as a result of ion etching [104]. They slightly decrease the effective width of strips in straight segments and increase the effective rounding radius of inner corners in bends. To describe within the GL approach, the reduction of the critical current in straight segments of the spiral, two identical defects were introduced at the opposite edges of model strip (Part A of the inset in Fig. 6.6). Both defects represent a local suppression of the order parameter in an area $\xi \times \xi$ adjacent to the strip edge. They reduce the critical current in the straight strip to 80% of the depairing critical current in the strip of the same width without defects. The dependence of the critical current in the strip with defects on the magnetic field is shown in Fig. 6.6 with open symbols. The critical current of the spiral, which is composed of bends and straight strips, will be limited to the smallest value of the critical currents of these two components.

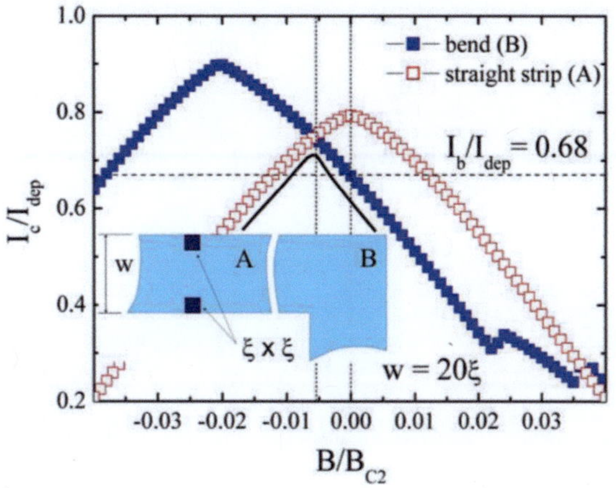

Figure 6.6: Relative critical current at different magnetic fields for a sharp bend in the strip with the width $W = 20\xi$ (closed symbols) and for the straight part of such a strip with defects (open symbols). The inset shows the geometry used for modeling: A, fragment of a straight strip with defects (not in scale); B, sharp bend. Solid line underlines currents that appear as critical currents of a spiral consisting from bends and straight strips. Vertical lines guide eyes to zero field and to the expected maximum in the critical current of the whole spiral.

The black solid line in Fig. 6.6 shows the path that the critical current of the model spiral should follow with a varying magnetic field. In accordance with experimental data, the maximum in the critical current occurs at ≈50 mT. It corresponds to the intersection of curves for the bend and for the straight strip. The experimental dependence of the critical current on the magnetic field can be modelled by a different set of rounding radii in the bend and the size of defects in the straight parts, e.g. smaller defects and a larger rounding radius. Since it's complicated to visualize defects, the set that was used to obtain the model currents is rather arbitrary. However, it does not anticipate any effects of this choice on the asymmetry in the rates of photon and dark counts in magnetic fields.

6.2.2 Dark and photon counts

The rate of dark counts in the square spiral is not symmetric with respect to the direction of either magnetic field or current. The minimum in the magnetic field dependence of the *DCR* appears for the same right field-current symmetry as the maximum in the magnetic field dependence of the critical current. This is illustrated in Fig. 6.7a, where the *DCR* is plotted as function of magnetic field for two opposite directions of bias current. Like the critical current, the *DCR* is invariant for changing simultaneously both field and current directions.

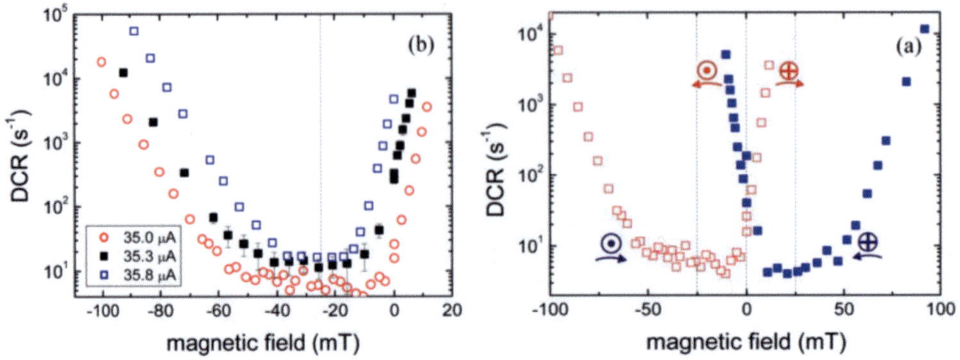

Figure 6.7: a) Rate of dark counts in magnetic field for two directions of the bias current with the magnitude 35 μA. The *DCR* for the positive current direction is shown with open symbols and for the negative direction with closed symbols. Pictograms depict combination of the field and current directions for each section of the plot. Dashed vertical lines are to guide the eyes; they show field positions of the minima in the *DCR* and zero field. b) Magnetic field dependencies of the *DCR* for different positive bias currents. Values of the bias current are specified in the legend. Vertical dashed line shows the location of the minimum on the field axis.

Noticeably, the minimum in *DCR* occurs at a field of approximately 25 mT, which is smaller than field corresponding to the maximum in the critical current. An increasing the bias current does not affect the position of the minimum in the *DCR* but makes it more pronounced and sharp (Fig. 6.7b). In circular spirals, the *PCR* and *DCR* were found to be symmetric with respect to magnetic field and current directions for any fields and currents (Fig. 6.8).

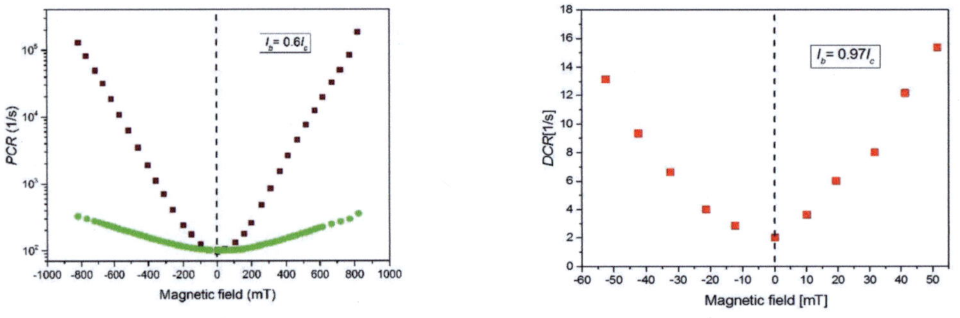

Figure 6.8: The dependence of *PCR* at 600 nm (green dots), 1500 nm (bro wn dots) (left graph) and *DCR* (right graph) on magnetic field of the circular spiral SNSPD biased at the current indicated in the legend.

Although, similar to the *DCR*, the rate of photon counts exhibits an asymmetry separately with respect to field and current directions, the effect of field and current appears more complicated [CSL+15]. First, the strength of asymmetry in the photon count rate (*PCR*) depends on the photon energy.

Figure 6.9 shows the rate of photon counts in magnetic fields for three wavelengths: 1400, 800, and 500 nm, and different bias currents. As reported earlier [44], the change in the *PCR*, which is produced by the same magnetic field, decreases with the decrease of the wavelength and varies from two orders of magnitude for the wavelength λ = 1400 nm to tens of a percent for λ = 500 nm. For each wavelength, an increase in bias current reduces the amount of *PCR* variation in the magnetic field. Although, like the *DCR*, *PCR* is also invariant for simultaneously changing the directions of both field and current, the asymmetry in the *PCR* qualitatively differs from the asymmetry in the *DCR*. Remarkably, the minimum in the *PCR* appears for the left field-current symmetry Fig. 6.10a and not for the right symmetry, as it is found for the *DCR*. In other words, for the same current direction, the minimum in the *PCR* is shifted in the opposite direction on the field axis as compared to the minimum in the *DCR*.

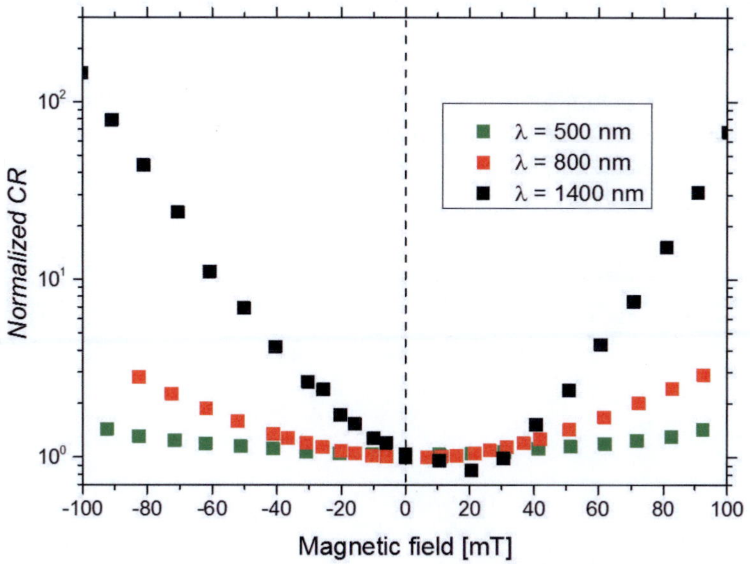

Figure 6.9: The dependences of rate of photon counts on magnetic field for square spiral SNSPD at different wavelengths: 500 (green dots), 800 (red dots) and 1400 nm (black dots).

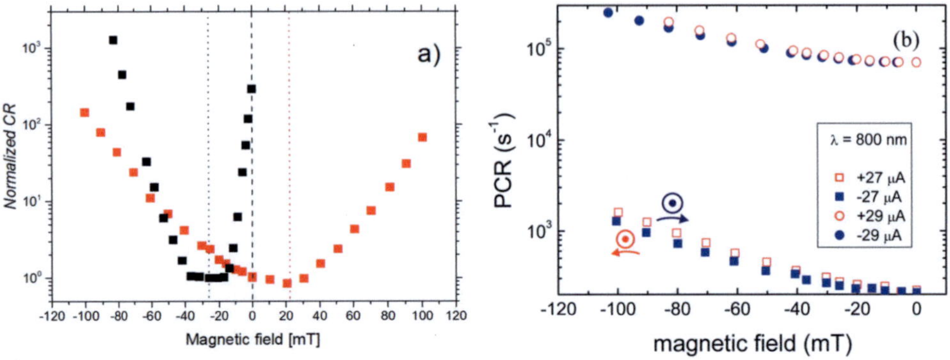

Figure 6.10: a) Rate of photon counts (red dots) measured at λ = 1400 nm and dark counts (black dots) in magnetic field of the square spiral SNSPD. Vertical dotted line marks the locations of the minimum *PCR*. b) Rate of photon counts of square spiral biased at different current indicated in legend in magnetic field for 800 nm wavelength.

For the wavelength 1400 nm at the field which corresponds to the minimum dark count rate (B = -25 mT), *PCR* becomes three times larger than *PCR* at B = 0 T. This leads to a five times enhancement of the single to noise ratio *SNR*(B) = *PCR*(B)/DCR(B) of the detector in a magnetic field (Fig. 6.11).

The *SNR* is a crucial parameter of the detector which determines the minimum detectable photon flux. The absolute values of the field at the *PCR* minima for λ = 1400 nm are approximately 17 mT, which is less than the absolute field values at the *DCR* minima.

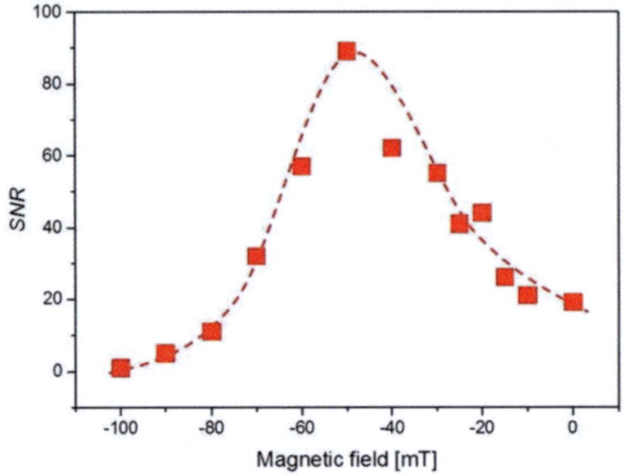

Figure 6.11: A single-to-noise ratio (*SNR*) of the square spiral SNSPD. The dashed line is to guide the eyes.

The asymmetry in the *PCR* is more pronounced for large wavelengths and small currents and disappears completely for wavelengths smaller than approximately 600 nm. For the wavelength 800 nm, the asymmetry is still distinguishable (Fig. 6.9). Within experimental accuracy, any asymmetry wasn't found for the wavelength of 500 nm (Fig. 6.9) as well as in the *PCR* dependences on the magnetic field for the circular spirals. The effect of an external magnetic field on the critical current and rates of dark and photon counts, which is described above, makes it possible to come to certain conclusions without invoking the qualitative microscopic analysis. Excluding a large single defect somewhere at the strip edge in the square spiral, it has to accept that any dark or photon count event, whose rate is asymmetric with respect to the direction of either field or current alone, comes from the bends in the spiral. For the critical current, the field effect was already demonstrated in experiments on single-bended strips [66]. The maximum of the experimental critical current in magnetic fields is achieved when critical current in the bend equals critical current in the straight parts of the strip. For count events in a straight strip, any microscopic model would predict symmetric field or current dependencies of corresponding rates because the strip itself and the absorption probability for photons are both symmetric over the midline of the strip, and the distributions of the current density and the magnetic field in the strip have even and odd symmetry, respectively, with respect to the midline. Hence, when the field or current direction changes, this will not affect the critical current and count rates, which should remain unchanged. The square spiral consists of straight strips and bends. Therefore, any asymmetry may come only from bends. Indeed, in circular spirals where no sharp corners are present, and the rounding radius of the spiral is much larger than the strip width, no asymmetry was observed. Furthermore, a weak asymmetry in the *DCR* with respect to the field direction has been observed in meanders [42] that contain turns with a different symmetry in small but non-equal numbers. Here, the net asymmetry may arise from a slight difference between geometrical shapes of individual turns. The phenomenological explanation of the asymmetry in the *DCR* is straightforward. For the left symmetry of the field current directions, the field increases the current density at the inner corner of each bend in the square spiral. This reduces the potential barrier for vortices entering the bend from the inner corner and, correspondingly, increases the rate of dark counts. The field applied in the right symmetry decreases the current density at the inner corner and decreases the count rate. When the field in the right symmetry further grows, the current density at the outer edges of straight strips increases and reduces the barrier for antivortices. At some field, they begin to dominate the net count rate, and the *DCR* starts to increase. Somewhere at an intermediate magnetic field, the *DCR* drops to a minimum. Since the rate of events from straight segments is symmetric with respect to the field direction and has a different field dependence as compared to the rate of events from bends, the net rate may have the minimum at a field smaller than the field that maximizes the critical current. Intuitively, one would expect the same kind of asymmetry for the rate of photon counts. However, this expectation silently postulates that photon and dark counts undergo the same microscopic scenario. This is not necessarily the case. Recently, it has been found that the microscopic scenario of photon detection as well as the detection efficiency may differ locally [24], [105], [106]. Let's consider the point on the common bisector of both corners in the bend close to its midline, e.g. point A in Fig. 6.4b. In the absence of an external field, the current density at the selected point is less than at the inner corner. The photon that is absorbed at this point creates a hotspot. The hollow in the order parameter forces the supercurrent to flow around and increases velocity of the condensate at the edges of the hotspot [105]. An external field in the left symmetry will decrease the current density locally around the hotspot. The photon is counted as a photon event if either the velocity locally reaches the critical value and a vortex-antivortex pair (VAP) appears or a vortex enters the hotspot from any side and then moves to the opposite one. Hence, an increasing field either disables VAP appearance or increases the barrier for the vortex around the hotspot. The local *PCR* decreases either way. Obviously, the field applied in the opposite direction causes an increase in the local *PCR*. The net effect crucially depends on the distribution of the photon absorption probability in the bend and on the bias current.

Model

Although dark counts are generated everywhere in the spiral, the rates per unit length (local *DCR*) depend crucially on the ratio between the local critical current and the bias current. The local *DCR* is proportional to $\exp(-\delta F/(k_B T))$ where δF is the local height of the barrier for vortex entry. In the framework of the London model for bias currents $I_b << I_C$, the barrier scales with the difference between the local critical current and the bias current $\delta F \propto \delta I = I_C(B) - I_b$ [100]. Therefore, for fields $B > 0.005 B_{c2}$, the local *DCR* in bends will be much higher than the local *DCR* in straight strips. The total *DCR* in the spiral will depend on the relative weight of bends and straight strips. However, since the local *DCR* in strips is symmetric with respect to the direction of the magnetic field, the presence of any asymmetry in the magnetic field dependence of the total *DCR* ensures the nonvanishing contribution of bends to the generation of dark counts. To quantitatively compare experimental results with the model calculations, the depairing current in the GL model is identified with the critical current in the straight strips in a zero magnetic field. Taking a 20% reduction of the model critical current in straight strips due to defects into account, it arrives at $I_b/I_C^{dep} = 0.68$ for the bias current, which is used for the *DCR* measurements. This relative bias current is marked with the straight dashed line in Fig. 6.6. If bends noticeably contribute to the total *DCR*, one would expect a minimum in the total *DCR* at a field close to experimental value of B_{max}. Plots in Fig. 6.7b confirm this expectation. There is a minimum in the *DCR* around −30 mT. Slopes of the *DCR* versus magnetic field are different for fields with right symmetry at $B < -50$ mT (fewer dark counts from bends) and for fields with left symmetry at $B > -10$ mT (more dark counts from bends). Different slopes correspond to different weights of bends and straight segments in the total *DCR*. Since the bends do not dominate in the total *DCR* at all magnetic fields, the minimum in the field dependence of the *DCR* does not coincide with the maximum in the $I_C(B)$ dependence. Accepting the thermally activated vortex crossing as the dominant photon-detection mechanism, one would expect for the *PCR* the same type of asymmetry as for the *DCR*. Indeed, the vortex should enter the superconductor and hotspot via the weakest place, i.e. the place where the current density/supervelocity is maximal. This can be the inner corner of the bend, especially when the hotspot is located close to it and the field of the left symmetry favors the entrance of a vortex with the same polarity as in the case of dark counts. However, contrary to dark count events, in the photon count scenario, the vortex should also exit the hotspot. The local value of the order parameter inside the hotspot Δ is less than the equilibrium value Δ_{eq} outside of the hotspot. If the relative local decrease of the order parameter is small $\delta = (\Delta_{eq} - \Delta)/\Delta_{eq} << 1$, the hotspot cannot pin the vortex, and it freely crosses the strip. When the relative decrease is large, the hotspot pins the vortex and prevents the photon count. Whichever of these two occurrences holds for a particular order parameter in the hotspot depends on the bias current.

In Fig. 6.12, the current at which the vortex leaves the hotspot is plotted as a function of the location of the hotspot in the bend. This current is called the detection current I_{det} since it ensures the photon count. Calculations are made in the framework of the modified hotspot model with the radius of the hotspot $R = 5\xi$ and the relative reduction of the order parameter $\delta = 0.4$. One can see that, close to the inner corner, there is an area in the bend where the small field of the left symmetry increases I_{det} while the field of the right symmetry decreases I_{det}. This area is marked schematically in gray in the inset of Fig. 6.12. Positions on the cut through this area at $y = W/4$ are circled in Fig. 6.12. Here, they span over the interval of bias currents $0.42I_{dep} < I_b < 0.49I_{dep}$. For any relative bias current within this interval, only the part of the circled area where $I_{det} < I_b$ provides photon counts. This active part decreases if a small positive (left symmetry) magnetic field is applied and increases if a magnetic field is negative (right symmetry). Because of the uniform and constant photon flux, the photon count rate increases with the increase in the area that is collecting photons. Hence, the active part will deliver the *PCR* with the asymmetry, which is observed experimentally. This asymmetry is inverted with respect to the "normal" asymmetry of the *DCR*. Note that, in the active part, the negative magnetic field favors exiting from the hotspot of the vortices that have entered the hotspot from the side of the inner corner. At the outer edge of the hotspot, far from the inner corner, the negative magnetic field increases the current density (and supervelocity) locally and decreases the energy barrier for vortex exit. The inverted asymmetry exists only in small magnetic fields. This corresponds to experimental observation. The *PCR* at $\lambda = 1400$ nm becomes symmetric for fields larger than 100 mT (Fig. 6.9). The inverted asymmetry disappears for $\delta = (\Delta_{eq} - \Delta)/\Delta_{eq} > 0.5$, which corresponds to photons with higher energy. In this case, I_{det} in the bend and in the straight strip are practically equal. The inverted asymmetry also disappears when the hotspot loses its ability to pin vortices, e.g. when $\delta < 0.3$. At bias currents larger than I_{det} in straight strips, photon counts come mostly from straight strips, and the *PCR* becomes symmetric. Under the same conventions as for dark counts, it was found that the current 27 μA, which is used to measure the *PCR* at $\lambda = 1400$ nm, corresponds to the model-relevant relative bias current $I_b = 0.5I_{dep}$. This current is at the upper boundary of the current interval where the inverted effect exists. However, since choice of the rounding radius and the size of defects for the model dependence of the critical current on the field is not unique, the relative bias current may have a different value. In other words, it's complicated, within the present model, to estimate numerically the relative weight of bends in the total rate of photon counts. The analysis [107] based on the solution of the time-dependent GL equations has shown that photon counts generated by bends differ from those originating from straight strips. More specifically, the overall duration of a *PCR* voltage pulse is smaller when the count comes from the bend and, at small bias currents, the amplitude of a *PCR* pulse from the bend is also smaller than the amplitude of the pulse from the straight strip. A paper based on the same theoretical approach [108] predicts a similar difference between the amplitudes of *PCR* voltage pulses originating from bends and straight strips with constrictions. With spirals, it was observed that *PCR* and *DCR* pulses with equal mean amplitudes and an amplitude spread, which is much narrower than both models predict. The time resolution of the present experiment (approximately 100 ps) does not allow to resolve the passage of kinematic vortices. Early experiments on meander structures, which include 180° turnarounds, had demonstrated a difference between mean amplitudes of *PCR* and *DCR* pulses as well as a decrease in the mean amplitude of *PCR* pulses with an increase of the photon energy [109]. These early observations contradict the results of both models. The reason for this discrepancy is not clear at this time. As in the case of dark counts, the asymmetry in the *PCR* itself ensures a noticeable contribution of the bends to the total rate of photon counts. It could be that the time-dependent GL equation alone cannot provide correct (quantitative) description of this problem.

Without solving coupled GL and kinetic equations, it is not possible to state unambiguously whether the passage of a single Abrikosov vortex or a series of kinematic vortices leaves enough heat to create a normal resistive domain. Instead of the kinetic equation, authors of the both models [107] solved the heat conductance equation. This approach is only qualitatively valid because the time for vortex nucleation is smaller than the electron-electron inelastic relaxation time, and the usage of the effective temperature is not justified. Furthermore, the local heating by a photon was reduced in the model of Ref. [107] by the choice of the coefficient, which describes heat transfer from electrons to phonons. Its value was larger than the typical value in NbN films. With a more realistic value for this coefficient [108], one finds that the photon absorbed near the bend creates a normal domain at a smaller current than that required for generating photon count in the straight strip.

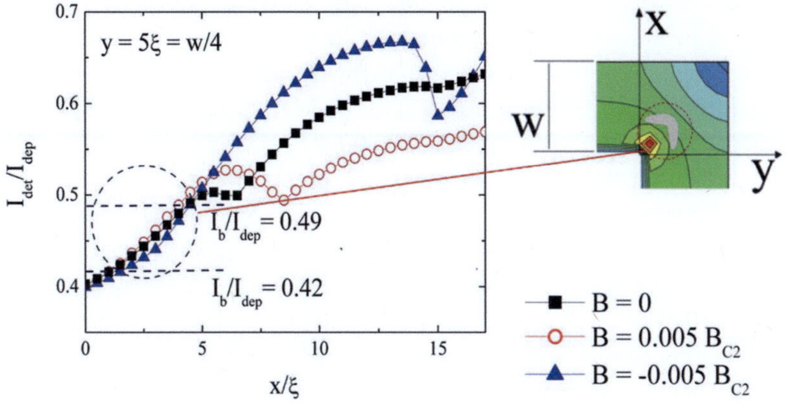

Figure 6.12: Detection current as a function of the positions (x) of the hotspot distances y = $W/4$ from the inner corner of the bend without magnetic field and for the magnetic field $B = 0.005B_{c2}$ with opposite directions. Positive magnetic field corresponds to the left symmetry. Coordinate system is shown in the inset in the panel. The inset in panel sketches the bend and the area (gray spot) where I_{det} is increased/decreased by small magnetic field of the left/right symmetry. The cut through this area at y = $W/4$ is marked with a blue dashed circle in the panel. Horizontal straight lines on graph show boundaries for bias currents within which the effect exists.

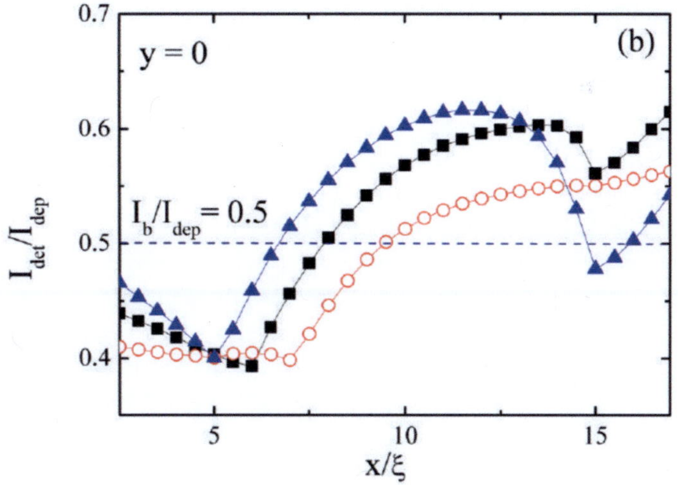

Figure 6.13: Detection current as a function of the positions (x) of the hotspot distances y = 0 from the inner corner of the bend without magnetic field and for the magnetic field $B = 0.005B_{c2}$ with opposite directions. Positive magnetic field corresponds to the left symmetry. Horizontal lines $I_b = 0.5I_{dep}$ show the nominal value of the bias current used to measure the *PCR* for the wavelength 1400 nm.

The theoretical considerations presented above are based on a relatively simple microscopic model of the hotspot [55]. The actual profile of the order parameter in the hotspot may differ from the assumption of this model that will quantitatively influence the pinning ability of a hotspot and the detection current. Furthermore, it's complicated precisely relate the bias current in the experiment to the particular relative bias current in the GL model. Therefore, the contribution to the net *PCR* from different parts of the bend remains largely undefined. For the bias current $I_b = 0.5I_{dep}$, the hotspot positions around the geometric border between the bend and the straight strip ($y = 0$, Fig. 6.13), which contribute to the net *PCR*, occupy an even larger area than the active positions in the central part of the bend. The positions at $0 < x/\xi < 5$ contribute with inverted asymmetry, whereas positions at $5 < x/\xi < 10$ contribute with normal asymmetry. The straight strips contribute symmetrically to the *PCR* at any bias current. Therefore, they smear out the shift of the minimum in the *PCR* to either side. The net *PCR* from all these areas may well be symmetric or show slight asymmetry. This interplay of photon counts from different parts of the spiral is further modified by the probability of photon absorption. This probability was computed for plane waves with three different polarizations. The probability of the photon absorption at a particular location is identified with the relative density of the high-frequency current, which is induced in the structure by the plane wave at normal incidence. Simulations were carried out with the software COMSOL [110], which uses the finite-element method. To verify that the simulation results were not affected by numerical instabilities or similar problems, the results obtained with COMSOL was compared to similar simulations done with the software Lumerical [111]. The latter is based on the finite-difference time-domain method. The results provided by these two techniques almost coincide. The COMSOL software solves numerically Maxwell's equations in the frequency domain. The spiral is modeled by its specific geometry and is represented by its frequency dependent dielectric function. The calculations lead to an accurate theoretical treatment of the problem, and the results automatically include surface plasmons if they are excited. Therefore, no further separate analysis of surface plasmons is necessary. In all simulations, a maximum mesh size of 7 nm and a complex dielectric function for NbN films in the normal state [23] were used. By comparing simulation results for a separate strip with a bend with the results for the whole spiral, it was confirmed that there was no crosstalk between adjacent strips via evanescent fields. The results are shown in Fig. 6.14 as grayscale plots.

They present relative current density in the equatorial surface of the bend at the frequency of the incident wave. For polarizations along the x or y axis (Fig. 6.14c), approximately half the bend is active in absorbing photons. The absorption probability is evenly distributed between hotspot positions, providing photon count rates with different asymmetries. The polarization at 45° is seen differently by adjacent bends. The two possibilities are shown in Figs. 6.14a and 6.14b.

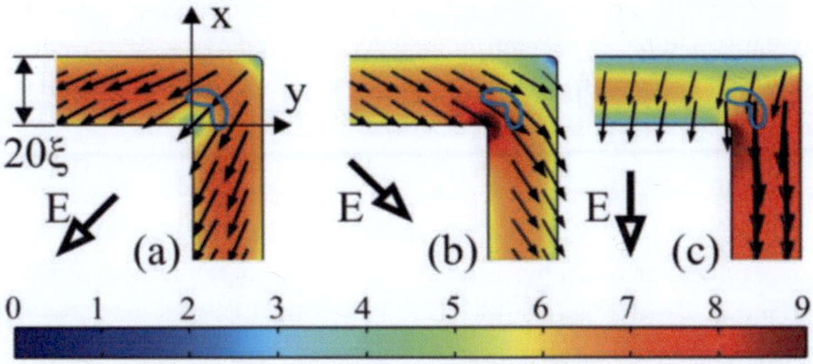

Figure 6.14: Relative probability of photon absorption in the bend and adjacent portions of the straight strips for the wavelength 1400 nm and different polarizations. Red (dark) color corresponds to the largest local probability. Blue curved lines circle the area that delivers the PCR with the inverted asymmetry. Black arrows show directions of currents that are excited in the bends by continuous electromagnetic waves. Polarization directions of the incident waves are shown with the empty arrows

The polarization along the bisector (Fig. Figs. 6.14a) delivers more photons to positions, providing normal asymmetry; whereas photons with perpendicular polarization (Fig. 6.14b) are more strongly absorbed at positions, providing inverted asymmetry. The net effect of the absorption probability on the asymmetry in the *PCR* seems to be very weak. In structures that include bends with single symmetry, the rates of photon and dark count events are asymmetric with respect to the direction of the external magnetic field and, separately, to the direction of the current. This asymmetry is associated with the asymmetry of the current crowding in the bends. Applying a simplified microscopic GL model, it has shown that count events, which provide asymmetry, come from bends while the rate of events coming from straight strips remains symmetric with respect to field and current directions. The microscopic scenario of the photon count event with intermediate pinning of the magnetic vortex in the hotspot explains the faint effect of the inverted asymmetry in the count rate for low-energy photons at small fields and currents. It has also shown that, at large magnetic fields and currents, the asymmetry in the rate of photon counts disappears as it is predicted by theoretical model.

6.2.3 Spectral detection efficiency

Although all models of single-photon detection of SNSPD (considered in chapter 2) predicts broadening of hot-spot spectrum with increase of the ratio I_C/I_C^{dep}, the effect of external magnetic field on detection of single photons is poor theoretically presented. No experimental verifications of this prediction have been reported. Obviously, the critical current of SNSPD deteriorate when external magnetic field is applied. It was experimentally demonstrated on SNSPD in form of meander [44] and nanodetector [112]. Due to the dual symmetry of meander layout and absence of symmetrical and asymmetrical turns in the nanodetector, the supercurrent cannot be enhanced by reduction of current crowding in bends. The specially designed square spiral layout with only one symmetrical turns is appropriate for applying of external magnetic field.

For an investigation of the magnetic-field dependence of critical currents, $I_C(B)$, magnetic fields up to 2 T were applied perpendicular to the spiral plane Fig. 6.15a. The measured critical current of detector made by negative-PMMA lithography with width of the nanowire $W = 90$ nm at $B = 0$ T is $I_C = 48.2$ μA. The critical current was also measured for both field and current directions as function of the magnetic field at $T = 4.2$ K. The square spirals revealed a dependence expected for single bends [66].

The measured critical current of the structure is reduced by current crowding [60] at the inner corners of bends with respect to the critical current of the straight parts. An external magnetic field induces screening current in bends. If screening current and bias current have opposite directions the field leads to a decrease of the local current density at the inner corners. This situation is demonstrated in Fig. 6.15b. The positive critical current (black dots) increases for negative magnetic fields, reaches a maximum at B_{max} = -56 mT and further decreases. The same effect was achieved when the directions of both field and current are changed to the opposite (Fig. 15b, red dots). The maximum of the critical current in magnetic field is found to be 12% higher than $I_C(0)$.

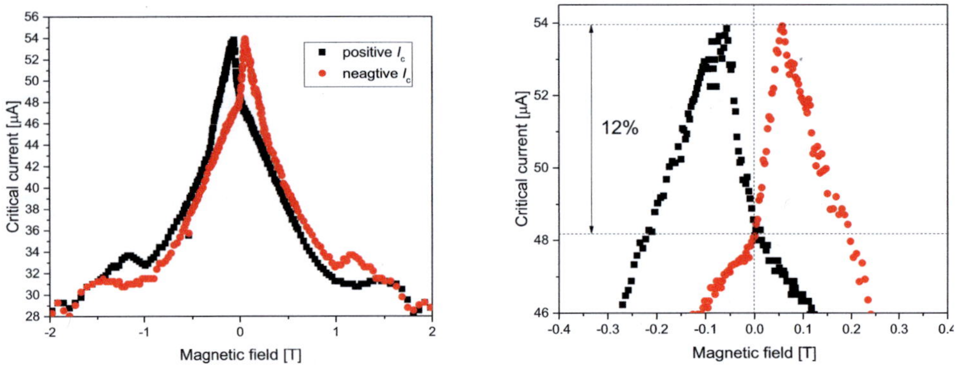

Figure 6.15: The dependence of the critical current on magnetic field of a square spiral SNSPD for positive (black symbols) and negative (red symbols) current directions.

In order to investigate the $\lambda_C(B)$ – dependence, the measurement of spectral detection efficiency from 400 up to 1100 nm wavelength has been done in magnetic fields. The detector was illuminated from the back side. In all measurements, the relative bias current was constant $0.95I_C(B)$. The detector has a cut-off wavelength of only 540 nm at 0 T with corresponds detection efficiency in the range of 12% (Fig. 6.16a). In the maximum I_C which corresponds to B_{max} = 56 mT, the detection efficiency grew up to 20 %. The hot-spot spectrum was broadened to 721 nm wavelength which corresponds 34% increase of cut-off wavelength. Although the critical current of detector is the same at B = 28 mT and B = 115 mT (Fig. 6.16b), the spectral cut-off wavelengths were found as 636 and 574 nm for smaller and higher magnetic field, respectively. The detection efficiency at these field is similar in the hot-spot region while the difference in *DE* is about 2 times at 1100 nm wavelength.

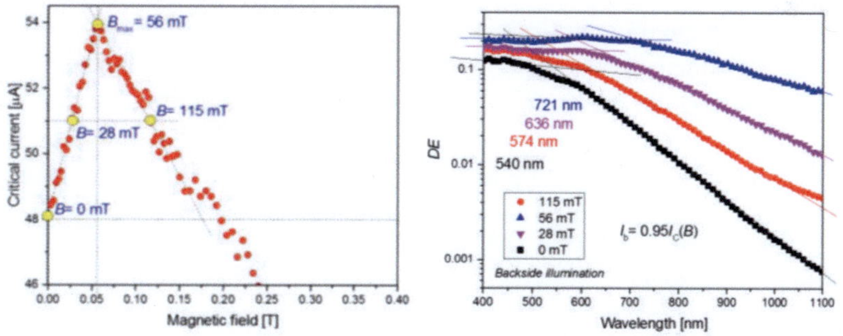

Figure 6.16: a) The dependence of the critical current on magnetic field of a square spiral SNSPD. The yellow points indicate the field chosen for measurement the spectral detection efficiency. b) The spectral detection efficiency of the square spiral SNSPD in the magnetic field indicated in the legend. The detector biased at the current $0.95I_C(B)$.

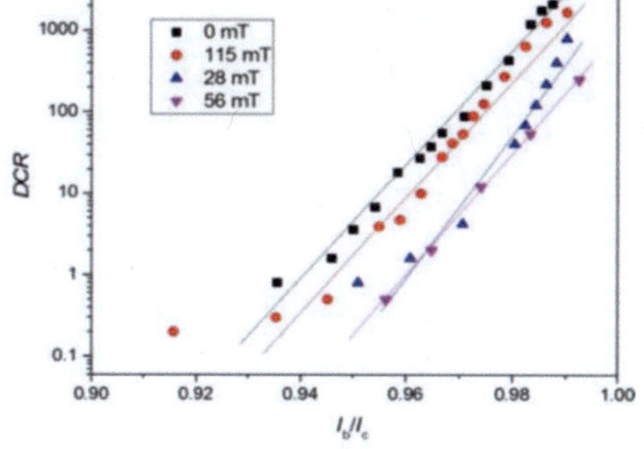

Figure 6.17: The dark count rate versus relative bias current presented for different magnetic fields indicated in the legend. The solid lines are to guide the eyes.

The event of dark counts was investigated in magnetic fields by measuring at bias current from 0.9 up to $0.99I_C(B)$. The minimum of dark counts was found at 56 mT, which corresponds the maximum $I_C(B)$. This result again confirms the assumption that, bends and turns with sharp corners are places from which dark counts most probably come. With applying a magnetic field, dark counts decrease till $DCR(B_{max})$. At $B = 115$ mT, the rate of the dark count is similar to $DCR(0\ \mathrm{T})$.

6.3 Summary

Using structures in the form of a square spiral, where all bends have the same symmetry with respect to the directions of bias current and external magnetic field, it was shown that areas around the bends largely contribute to the rate of dark counts and to the rate of photon counts at small photon energies. The minimum detectable photon flux was decreased by applying external magnetic field to the detector. The minimum in the rate of dark counts reproduces the asymmetry of the maximum in the critical current as a function of the magnetic field. Contrarily, the minimum in the rate of photon counts demonstrates opposite asymmetry. The asymmetry of *PCR* and *DCR* is associated with the asymmetry of the current crowding in the bends. Invoking handedness of the observed asymmetries and mapping the computed local absorption probability for photons and the local detection threshold current, areas in the bends where, at low photon energies, photon counts occur is identified. The microscopic scenario of the photon count event explains the faint effect of the inverted asymmetry in the count rate for low-energy photons at small fields and currents. At large magnetic fields and currents, the asymmetry in the rate of photon counts disappears as it is predicted by the theoretical model. The asymmetry in count rates is absent in circular spirals without bends.

The demonstrated effect of extending the hot-spot response spectrum by applying a magnetic field can be used in applications for quantum-photonic circuits. The detection efficiency of a nanowire with a single bend on top of a diamond photonic waveguide can be enhanced by an external magnetic field.

7 Conclusions

In this thesis, approaches for the improvement of the spectral bandwidth of Superconducting Nanowire Single-Photon Detectors (SNSPD) have been studied. Nowadays, the device technology based on superconducting nanowire is a highly promising alternative to avalanche photodiodes in the area of single-photon detection. In this thesis, main investigations are focused to extend the spectral bandwidth and enhancement of detection efficiency, especially in infrared range. The detection efficiency is limited by the absorbance of NbN film (only ~30%). The absorbance is enhanced by variation of stoichiometry, optical coupling with backside illumination of NbN one-side-polished sapphire. The absorbance of a thick 5 nm NbN film is reached 50% in absolute value without using any additional cavity or mirror.

The spectral bandwidth of the detection efficiency is restricted by an exponential decay once the cut-off wavelength is surpassed. The cut-off wavelength is partially shifted by optimization of material properties of the superconductor. According to the theoretical models, to improve the spectral bandwidth of an SNSPD, the ratio of the bias current to the depairing critical current has to be increased. The maximum experimentally achievable critical current in an SNSPD usually stays well below the theoretical limit of the pair-breaking current.

The first reason is the suppression of the critical current due to the proximity effect in NbN film. In this thesis, a detailed and systematic investigation of the proximity effect is performed for NbN thin film strips in a wide range of the width. The proximity effect is reduced in bi-layer NbN/AlN in the comparison to single-layer NbN. The *in-situ* deposited NbN/AlN layer has up 1 K higher critical temperature than single-layer NbN with the same nominal thickness below 6 nm. It allows to reduce the cross-section of nanowires for improvement of the sensitivity to low-energy photons.

The second reason of low ratio of experimental to theoretical currents is rough edges of the nanowire which causes a larger number of defects and correspondingly a reduction in the experimental critical current. The usage of the negative-PMMA lithography leads to an enhancement of both superconductivity in strips and their cooling efficiency to substrate due to smaller roughness of edges. The negative-PMMA lithography offers the following advantages for fabrication of SNSPDs over the positive-PMMA lithography. In electron-beam lithography, smaller thickness of the resist layer, in general, allows for writing smaller pixels in a more reproducible manner. Therefore, further reduction of the width of nanowires is possible with negative-PMMA process while keeping the protection properties of the resist at high enough level. While the enhancement of the critical temperature of negative-PMMA nanowires is about 0.5 K, which is only 5% of the T_C of the positive-PMMA strips, the increase in the critical current density measured at 4.2 K is much more significant. The latter makes the negative-PMMA technology even more attractive for optimization of SNSPDs. The higher critical temperature of the negative-PMMA nanowires allows further decreasing the cross-section of the nanowire while keeping T_C high enough for operation of SNSPD at 4.2 K. Larger critical current, which can be realized for a given width, means larger signal-to-noise ratio (SNR) in the voltage-pulse response of the detector. A higher T_C corresponds to a larger energy gap. With other material parameter and operation conditions being equal, this should reduce the value of the cut-off wavelength. From the other side, thinner films with lower critical temperature (the critical temperature decreases with decreasing thickness due to the proximity effect) can be used to fabricate SNSPD for operation at a given temperature, e.g. at $T = 4.2$ K. Furthermore, higher T_C is one of the reasons for higher j_C of the negative-PMMA nanowires. However, an increase in T_C alone cannot explain the almost 40% increase in j_C(4.2 K). The j_C/j_C^{dep}(4.2K) ratio for the negative-PMMA strips is approximately 0.67 (± 3) that is noticeably larger than the value 0.48 (± 5) obtained for the positive-PMMA strips. According to the theoretical models of the SNSPD response an increase of this ratio should result in the shift of the cut-off wavelength towards longer wavelengths.

Although the superconductivity in the negative-PMMA strips is enhanced the ratio j_C/j_C^{dep}(4.2 K) is still less than one. One of the possible reasons for that can be the particular stoichiometry of NbN films. The films were deposited at discharge parameters providing maximum of T_C. Larger j_C/j_C^{dep}(4.2 K) ratio is achieved in films with a larger relative content of Nb which, however, have slightly smaller critical temperature. Another reason is revealed by the temperature dependence of the critical current of a nanowire. In the $I_C(T)$-dependence, local density of the critical current in the superconducting core of narrow strips approaches the density of the Ginzburg-Landau depairing current. The strips with $W > 200$ nm demonstrate flattening of $I_C(T)$ at low temperatures. The flattening is stronger and begins at higher temperatures (closer to T_C) for wider strips. The critical currents are too low to generate a magnetic field above the penetration field, B_{stop}. The mechanisms of this deviation and of the flattening of the temperature dependencies of critical current are not fully understood yet.

The third reason of suppressed experimental current is the sharp turns of the nanowire of detectors usually made in the form of a meander line. To overcome this issue, a single-spiral layout was proposed as alternative geometry for SNSPD. The single-spiral technology was developed for two types of detectors: Archimedean spiral and square spiral. All technological steps are optimized to minimize the influence of the process of fabrication of detectors on the superconducting properties of the films. SNSPDs shaped as either meander or spirals are systematically characterized and compared. The spirals had different layouts, a double-spiral layout with an S-turn in the middle and a single-spiral layout without such turn. For detectors of each layout, the spectra of the single-photon response were measured in the wavelength range from 400 nm to 1600 nm and defined the cut-off wavelength (λ_C) beyond which the response rolls off. The largest and the smallest λ_c were found for the single-spiral layout and for the meander, respectively. For all three layouts the relationship between λ_C and the relative bias current falls onto a universal curve which has been predicted in the framework of the modified hot-spot model. For the single-spiral detector, the efficiency of photon detection at wavelengths smaller than λ_C reaches the expected absorbance of the spiral structure and the timing jitter per unit length of the nanowire has the smallest value. In terms of the detection efficiency and spectral cut-off the most important scaling factor is the critical current for the particular layout while the bends play a minor role. Contrary, the rate of dark counts as well as the timing jitter are mostly restricted by the quality and number of bends in the layout. The value of timing jitter is directly related to the length of the nanowire, its kinetic inductance and the amount of bends.

The alternative approach for improving the spectral bandwidth is to apply an external magnetic field to detectors in the form of a square spiral, where all bends have the same symmetry with respect to the directions of the bias current and external magnetic field. The minimum detectable photon flux is decreased by applying external magnetic field to the detector. The minimum in rate of dark counts reproduces the asymmetry of the maximum in critical current as a function of the magnetic field. Contrarily, the minimum in the rate of photon counts demonstrates the opposite asymmetry. The asymmetry of *PCR* and *DCR* is associated with the asymmetry of the current crowding in the bends. Invoking handedness of the observed asymmetries and mapping the computed local absorption probability for photons and the local detection threshold current, areas in the bends where, at low photon energies, photon counts occur is identified. The microscopic scenario of the photon count event explains the faint effect of the inverted asymmetry in the count rate for low-energy photons at small fields and currents. At large magnetic fields and currents, the asymmetry in the rate of photon counts disappears as it is predicted by normal-core vortex model.

The proposed approaches for improvement of performance of single-photon detectors open new doors for a wide range of applications. The developed techniques in frame of this PhD work offer a range of advantages for applications of classical and quantum optics.

Bibliography

[1] C. Natarajan, M. Tanner, and R. Hadfield, Superconducting nanowire single-photon detectors: physics and applications, Supercond. Sci. Technol. 25, 063001, 2012.

[2] A. Tada, N. Namekata, S. Inoue, Sinusoidally gated InGaAs/InP avalanche photodiode with 53% photon detection efficiency at 1550 nm, Lasers and Electro-Optics (CLEO), 2016.

[3] E. Amri, G. Boso, B. Korzh, and H. Zbinden, Temporal jitter in free-running InGaAs/InP single-photon avalanche detectors, Optics Letters Vol. 41, Issue 24, pp. 5728-5731, 2016.

[4] M. A. Itzler, U. Krishnamachari, Q. Chau, X. D. Jiang, M. Entwistle, M. Owens, and K. Slomkowski, Statistical analysis of dark count rate in Geiger-mode APD FPAs, Proc. SPIE, vol. 9250, 2014.

[5] J. Zhang, M. A Itzler, H. Zbinden and J.-W. Pan, Advances in InGaAs/InP single-photon detector systems for quantum communication, Science & Applications 4, 2015.

[6] M. A. Moreno, Next generation avalanche photodiodes: realising new potentials, PhD thesis, The University of Sheffield, 2016.

[7] A. D. Semenov, G. N. Gol'tsman, and A. A. Korneev, Quantum detection by current carrying superconducting film, Physica C: Superconductivity, 351, 349 – 356, 2001.

[8] S. N. Dorenbos, E. M. Reiger, U. Perinetti, V. Zwiller, T. Zijlstra, and T. M. Klapwijk, High performance fiber-coupled NbTiN superconducting nanowire single photon detectors with Gifford-McMahon cryocooler, Appl. Phys. Lett. 93(13), 131101, 2008.

[9] F. Marsili, V. B. Verma, J. A. Stern, S. Harrington, A. E. Lita, T. Gerrits, I. Vayshenker, B. Baek, M. D. Shaw, R. P. Mirin, and S. W. Nam, Detecting single infrared photons with 93% system efficiency, Nature Photonics 7, 210–214, 2013.

[10] S. N. Dorenbos, P. Forn-Diaz, T. Fuse, A. H. Verbruggen, T. Zijlstra, T. M. Klapwijk, and V. Zwiller, Low gap superconducting single photon detectors for infrared sensitivity, Appl. Phys. Lett. 98, 251102, 2011.

[11] V. B. Verma, A. E. Lita, M. R. Vissers, F. Marsili, D. P. Pappas, R. P. Mirin, and S. W. Nam, Superconducting nanowire single photon detectors fabricated from an amorphous Mo0.75Ge0.25 thin film, Appl. Phys. Lett.105, 022602, 2014.

[12] Yu. Korneeva, M. Mikhailov, Yu. Pershin, N. Manova, A. Divochiy, Yu. Vakhtomin, A. Korneev, K., Superconducting single-photon detector made of MoSi film, Supercond. Sci. Technol. 27, 095012, 2014.

[13] F. Marsili, F. Bellei, F. Najafi, A. E. Dane, E. A. Dauler, R. J. Molnar, K. K. Berggren, Efficient single photon detection from 500 nm to 5 μm wavelength, Nano letters 12 (9), 4799-4804, 2012.

[14] V. Shcheslavskiy, P. Morozov, A. Divochiy, Yu. Vakhtomin, K. Smirnov, and W. Becker, Ultrafast time measurements by time-correlated single photon counting coupled with superconducting single photon detector, Rev. Sci. Instrum. 87, 053117, 2016.

[15] F. H. L. Koppens, T. Mueller, Ph. Avouris, A. C. Ferrari, M. S. Vitiello, and M. Polini, Photodetectors based on graphene, other two-dimensional materials and hybrid systems, Nature Nanotechnology 9, 780–793, 2014.

[16] D. Henrich, S. Dörner, M. Hofherr, K. Il'in, A. Semenov, E. Heintze, M. Scheffler, M. Dressel, M. Siegel, Broadening of hot-spot response spectrum of superconducting NbN nanowire single-photon detector with reduced nitrogen content, J. Appl. Phys. 112, 074511, 2012.

[17] D. H. Andrews, R. D. Fowler, and M. C. Williams, The Effect of Alpha-particles on a Superconductor, Phys. Rev. 76, 154, 1949.

[18] D. E. Spiel, R. W. Boom, and E. C. Crittenden, Thermal spikes in superconducting thin films of Sn and In, Applied Physics Letters 7 (11), 292-294, 1965.

[19] R.F. Broom and E.H. Rhoderick, Thermal propagation of a normal region in a thin superconducting film and its application to a new type of bistable element, British Journal of Applied Physics 11 (7), 292-296, 1960.

[20] K. E. Gray, R. T. Kampwirth, J. F. Zasadzinski, and S. P. Ducharmet, Thermal propagation and stability in superconducting films, Journal of Physics F: Metal Physics 13 (2), 405-430, 1983.

[21] G. Gol'tsman, O. Okunev, G. Chulkova, A. Lipatov, A. Semenov, K. Smirnov, B. Voronov, A. Dzardanov, C. Williams, and R. Sobolewski, Picosecond superconducting single-photon optical detector, Appl. Phys. Lett. 79, 705-707, 2001.

[22] K. Tranthama and T. J. Reece, Demonstration of the Airy disk using photography and simple light sources, American Journal of Physics 83, 928, 2015.

[23] A. Semenov, B. Günther, U. Böttger, H.-W. Hübers, H. Bartolf, A. Engel, A. Schilling, K. Ilin and M. Siegel, R. Schneider and D. Gerthsen, and N. A. Gippius, Optical and transport properties of ultrathin NbN films and nanostructures, Physical Review B 80, 054510, 2009.

[24] J. J. Renema, Q. Wang, R. Gaudio, I. Komen, K. op't Hoog, D. Sahin, A. Schilling, M.P. van Exter, A. Fiore, A. Engel, and M.J.A. de Dood, Position-Dependent Local Detection Efficiency in a Nanowire Superconducting Single-Photon Detector, Nano Letters 15, 4541−4545, 2015.

[25] W. J. Skocpol, M. R. Beasley, and M. Tinkham, Self-heating hotspots in superconducting thin-film microbridges, Journal of Applied Physics 45 (9), 4054-4066, 1974.

[26] A. Vl. Gurevich and R. G. Mints, Self-heating in normal metals and superconductors, Reviews of Modern Physics 59 (4), 941-999, 1987.

[27] A. D. Semenov, P. Haas, B. Günther, H.-W. Hübers, K. Il'in, M. Siegel, A. Kirste, J. Beyer, D. Drung, T. Schurig, and A. Smirnov, An energy-resolving superconducting nanowire photon counter, Superconductor Science and Technology 20, 919–924, 2007.

[28] A Engel, J. J. Renema, K. Il'in, and A. Semenov, Detection mechanism of superconducting nanowire single-photon detectors, Supercond. Sci. Technol. 28, 114003, 2015.

[29] B. Baek, A. E. Lita, V. Verma, and S. W. Nam, Superconducting a-WxSi1-x nanowire single-photon detector with saturated internal quantum efficiency from visible to 1850 nm, Applied Physics Letters 98, 251105, 2011.

[30] H. Li, W. Zhang, L. You, L. Zhang, X. Yang, X. Liu, S. Chen, C. Lv, W. Peng, Z. Wang, and X. Xie, Nonideal Optical Cavity Structure of Superconducting Nanowire Single-Photon Detector, IEEE Transactions on Applied Superconductivity 6, 3803705, 2014.

[31] R. Lusche, A. Semenov, K. Il'in, Y. Korneeva, A. Trifonov, A. Korneev, H.-W. Hübers, M. Siegel, and G. Gol'tsman, Effect of the Wire Width and Magnetic Field on the Intrinsic Detection Efficiency of Superconducting Nanowire Single-Photon Detectors, IEEE Transactions on Applied Superconductivity 23 (3), 2200205, 2012.

[32] M. Hofherr, D. Rall, K. Ilin, M. Siegel, A. Semenov, H.-W. Hübers, and N. A. Gippius, Intrinsic detection efficiency of superconducting nanowire single-photon detectors with different thicknesses, Journal of Applied Physics 108, 014507, 2010.

[33] A. Engel, K. Inderbitzin, A. Schilling, R. Lusche, A. Semenov, H.-W. Hübers, D. Henrich, M. Hofherr, K. Il'in, and M. Siegel, Temperature-Dependence of Detection Efficiency in NbN and TaN SNSPD, IEEE Transactions on Applied Superconductivity 23 (3), 2300505, 2013.

[34] A. Semenov, private communication.

[35] A. Engel, A. Aeschbacher, K. Inderbitzin, A. Schilling, K. Il'in, M. Hofherr, M. Siegel, A. Semenov, and H.-W. Hübers, Tantalum nitride superconducting single-photon detectors with low cut-off energy, Appl. Phys. Lett. 100, 062601, 2012.

[36] D. Henrich, L. Rehm, S. Dörner, M. Hofherr, K. Il'in, A. Semenov, M. Siegel, Detection Efficiency of a Spiral-Nanowire Superconducting Single-Photon Detector, IEEE Trans. Applied Superconductivity 23, 2200405, 2013.

[37] S. N. Dorenbos, E. M. Reiger, N. Akopian, U. Perinetti, V. Zwiller, T. Zijlstra, and T. M. Klapwijk, Superconducting single photon detectors with minimized polarization dependence, Appl. Phys. Lett. 93, 161102, 2008.

[38] A. Annunziata, D. Santavicca, J. Chudow, L. Frunzio, M. Rooks, A. Frydman, and D. Prober, Niobium Superconducting Nanowire Single-Photon Detectors, IEEE Transactions on Applied Superconductivity 19, 327, 2009.

[39] K. S. Il'in, M. Lindgren, M. Currie, A. D. Semenov, G. N. Gol'tsman, R. Sobolewski, S. I. Cherednichenko, and E. M. Gershenzon, Picosecond hot-electron energy relaxation in NbN superconducting photodetectors, Applied Physics Letters, 76, 2752–2754, 2000.

[40] F. Marsili, D. P. Cunnane, R. M. Briggs, A. D. Beyer, M. D. Shaw, B. S. Karasik, M. A. Wolak, N. Acharya, X. X. Xi, Superconducting nanowire detectors based on MgB2, IEEE Lasers and Electro-Optics (CLEO), 1-2, 2015.

[41] D. Yu. Vodolazov, Single-Photon Detection by a Dirty Current-Carrying Superconducting Strip Based on the Kinetic-Equation Approach, Phys. Rev. Applied 7, 2017.

[42] A. Engel, A. Schilling, K. Il'in, and M. Siegel, Dependence of count rate on magnetic field in superconducting thin-film TaN single-photon detectors, Phys. Rev. B 86, 140506(R), 2012.

[43] M. K. Akhlaghi, H. Atikian, A. Eftekharian, M. Loncar, and A. H. Majedi, Reduced dark counts in optimized geometries for superconducting nanowire single photon detectors, Optics Express Vol. 20, Issue 21, pp. 23610-23616, 2012.

[44] R. Lusche, A. Semenov, Y. Korneeva, A. Trifonov, A. Korneev, G. Gol'tsman, and H.-W. Hübers, Effect of magnetic field on the photon detection in thin superconducting meander structures, Phys. Rev. B 89, 104513, 2014.

[45] M. V. Sidorova, A. V. Divochiy, Y. B. Vakhtomin, and K. V. Smirnov, Ultrafast superconducting single-photon detector with a reduced active area coupled to a tapered lensed single-mode fiber, Journal of Nanophotonics 9(1), 093051, 2015.

[46] Q. Zhao, L. Zhang, T. Jia, L. Kang, W. Xu, J. Chen, and P. Wu, Intrinsic timing jitter of superconducting nanowire single-photon detectors, Appl. Phys. B 104, 673–678, 2011.

[47] J. A. O'Connor, M. G. Tanner, C. M. Natarajan, G. S. Buller, R. J. Warburton, S. Miki, Z. Wang, S. W. Nam, and R. H. Hadfield, Spatial dependence of output pulse delay in a niobium nitride nanowire superconducting single-photon detector, Applied Physics Letters 98, 201116, 2011.

[48] A. G. Kozorezov, C. Lambert, F. Marsili, M. J. Stevens, V. B. Verma, J. P. Allmaras, M. D. Shaw, R. P. Mirin, and S. W. Nam, Fano fluctuations in superconducting nanowire single-photon detectors, Phys. Rev. B 96, 054507, 2017.

[49] J. Zhang, W. Słysz, A. Pearlman, A. Verevkin, R. Sobolewski, O. Okunev, G. Chulkova, and G. N. Gol'tsman,, Time delay of resistive-state formation in superconducting stripes excited by single optical photons, Physical Review B 67, 132508, 2003.

[50] K. Harrabi and J. P. Maneval, Measurements of the delay time between a critical current pulse and the first resistive response in superconducting niobium strips, IEEE Transactions on Applied Superconductivity 27 (4), 2201104, 2017.

[51] H. Wu, C. Gu, Y. Cheng, and Xi. Hu, Vortex-crossing-induced timing jitter of superconducting nanowire single-photon detectors, Applied Physics Letters 111, 062603, 2017.

[52] L. N. Bulaevskii, M. J. Graf, C. D. Batista, and V. G. Kogan, Vortex-induced dissipation in narrow current-biased thin-film superconducting strips, Phys. Rev. B 83, 144526, 2011.

[53] J. J. Renema, R. Gaudio, Q. Wang, Z. Zhou, A. Gaggero, F. Mattioli, R. Leoni, D. Sahin, M. J. A. de Dood, A. Fiore, and M. P. van Exter, Experimental Test of Theories of the Detection Mechanism in a Nanowire Superconducting Single Photon Detector, Physical Review Letters.

[54] Q. Wang, J. J. Renema, A. Engel, M. P. van Exter, and M. J. A. de Dood, Local detection efficiency of a NbN superconducting single photon detector explored by a scattering scanning near-field optical microscope, Optics Express 23(19), 24873, 2015.

[55] D. Yu. Vodolazov, Yu. P. Korneeva, A. V. Semenov, A. A. Korneev, and G. N. Goltsman, Vortex-assisted mechanism of photon counting in a superconducting nanowire single-photon detector revealed by external magnetic field, Phys. Rev. B 92, 104503, 2015.

[56] V. P. Andratskii, L. M. Grundel, V. N. Gubankov, and N. B. Pavlov, Destruction of superconductivity in thin narrow films by a current, Sov. Phys. JETP 38, 797, 1974.

[57] A. Yu. Rusanov, M. B. S. Hesselberth, and J. Aarts, Depairing currents in superconducting films of Nb and amorphous MoGe, Phys. Rev. B 70, 024510, 2004.

[58] S. Nawaz, R. Arpaia, F. Lombardi, and T. Bauch, Microwave Response of Superconducting YBa2Cu3O7−δ Nanowire Bridges Sustaining the Critical Depairing Current: Evidence of Josephson-like Behavior, Physical Review Letters 110, 167004, 2013.

[59] K. K. Likharev, Izv. Vuzov.Radiofizika,14, 909, 1971.

[60] J. R. Clem and K. K. Berggren, Geometry-dependent critical currents in superconducting nanocircuits, Physical Review B 84, 174510, 2011.

[61] D. Henrich, P. Reichensperger, M. Hofherr, J. M. Meckbach, K. Il'in, M. Siegel, A. Semenov, A. Zotova, D. Yu. Vodolazov, Geometry-induced reduction of the critical current in superconducting nanowires, Physical Review B 86,144504, 2012.

[62] H. L. Hortensius, E. F. C. Driessen, T. M. Klapwijk, K. K. Berggren, and J. R. Clem, Critical-current reduction in thin superconducting wires due to current crowding, Applied Physics Letters 100, 182602, 2012.

[63] O. A. Adami, D. Cerbu, D. Cabosart, M. Motta, J. Cuppens, W. A. Ortiz, V. V. Moshchalkov, B. Hackens, R. Delamare, J. Van de Vondel, A. V. Silhanek, Current crowding effects in superconducting corner-shaped Al microstrips, Applied Physics Letters 102, 052603, 2013.

[64] A. M. Goldman, Superconductor insulator transition, Int. J. Mod. Phys. B, 24, 4081, 2010.

[65] J. R. Clem, Y. Mawatari, G. R. Berdiyorov, and F. M. Peeters, Predicted field-dependent increase of critical currents in asymmetric superconducting nanocircuits, Phys. Rev. B 85, 144511, 2012.

[66] K. Ilin, D. Henrich, Y. Luck, Y. Liang, M. Siegel, and D. Yu. Vodolazov, Critical current of Nb, NbN, and TaN thin-film bridges with and without geometrical nonuniformities in a magnetic field, Phys. Rev. B 89, 184511, 2014.

[67] S. A. Wolf, J. J. Kennedy, and M. Nisenoff, Properties of superconducting rf sputtered ultrathin films of Nb, Journal of Vacuum Science and Technology, 13, 145–147, 1976.

[68] S. Bechstein, C. Köhn, D. Drung, J.-H. Storm, O. Kieler, V. Morosh, and T. Schurig, Investigation of nanoSQUID designs for practical applications, Supercond. Sci. Technol. 30, 034007, 2017.

[69] J. L. Kloosterman, D. J. Hayton, Y. Ren, T. Y. Kao, J. N. Hovenier, J. R. Gao, T. M. Klapwijk, Q. Hu, C. K. Walker, and J. L. Reno, Hot electron bolometer heterodyne receiver with a 4.7-THz quantum cascade laser as a local oscillator, Appl. Phys. Lett. 102, 011123, 2013.

[70] J. M. Meckbach, M. Merker, S. J. Buehler, K. Ilin, B. Neumeier, U. Kienzle, E. Goldobin, R. Kleiner, D. Koelle, M. Siegel, Sub-μm Josephson Junctions for Superconducting Quantum Devices, IEEE Transactions on Applied Superconductivity 23 (3), 2013.

[71] Y. V. Fominov and M. V. Feigel'man, Superconductive properties of thin dirty superconductor–normal-metal bilayers, Physical Review B 63, 094518, 2001.

[72] D. Henrich, Influence of Material and Geometry on the Performance of Superconducting Nanowire, PhD thesis, Karlsruhe Insitute of Technology, 2013.

[73] N. R. Werthamer, E. Helfand, and P. C. Hohenberg, Temperature and Purity Dependence of the Superconducting Critical Field, Physical Review 147, 295, 1966.

[74] U. S. Pracht, E. Heintze, C. Clauss, D. Hafner, R. Bek, D. Werner, S. Gelhorn, M. Scheffler, M. Dressel, D. Sherman, B. Gorshunov, K. S. Il'in, D. Henrich, and M. Siegel, Electrodynamics of the Superconducting State in Ultra-Thin Films at THz Frequencies, IEEE Transactions on Terahertz Science and Technology 3, 269, 2013.

[75] T. Sun, Classical Size Effect In Copper Thin Films: Impact Of Surface And Grain Boundary Scattering On Resistivity, PhD thesis, University of Central Florida, 2009.

[76] H. Duan, J. Zhao, Y. Zhang, E. Xie and L. Han, Preparing patterned carbonaceous nanostructures directly by overexposure of PMMA using electron-beam lithography, Nanotechnology 20, 135306, 2009.

[77] M. Bowden, Materials for Microlithography, ACS Symp. Series 266, pp 39–123, 1984.

[78] A. Hoole, M. Welland, and A. Broers, Negative PMMA as a high-resolution resist - the limits and possibilities, Semicond. Sci. Technol. 12, 1166–1170, 1997.

[79] I. Zailer, J. Frost, V. Chabasseur-Molyneux, C. Ford, and M. Pepper, Crosslinked PMMA as a high-resolution negative resist for electron beam lithography and applications for physics of low-dimensional structures, Semicond. Sci. Technol. 11, 1235–1238, 1996.

[80] W. Rong, Z. Li, W. Zhang, L. Sun, An improved Canny edge detection algorithm, IEEE International Conference on Mechatronics and Automation, 14546930, 2014.

[81] L. Aslamazov, and A. Larkin, The influence of fluctuation pairing of electrons on the conductivity of normal metal, Phys. Lett. 26A, 238, 1968.

[82] K. Ilin, D. Rall, M. Siegel, A. Engel, A. Schilling, A. Semenov, H.-W. Huebers, Influence of thickness, width and temperature on critical current density of Nb thin film structures, Physica C 470, 953, 2010.

[83] K. Ilin, D. Rall, M. Siegel, A. Semenov, Critical current density in thin superconducting TaN film structures, Physica C 479, 176, 2012.

[84] A. Stockhausen, K. Ilin, M. Siegel, U. Soedervall, P. Jedrasik, A. Semenov, H.-W. Huebers, Adjustment of self-heating in long superconducting thin film NbN microbridges, Supercond. Sci. Technol. 25, 035012, 2012.

[85] S. Sarfraz, Influence of developer and proximity effect on morphology of edge roughness in lithographically graded exposures using PMMA, IEEE International Multitopic Conference, 10471355, 2008.

[86] G. M. Maksimova, N. V. Zhelezina and I. L. Maksimov, Critical current and negative magnetoresistance of superconducting film with edge barrier, Europhys. Lett., 53 (5), pp. 639–645, 2001.

[87] G. R. Boogaard, A. H. Verbruggen, W. Belzig, and T. M. Klapwijk, Resistance of superconducting nanowires connected to normal-metal leads, Physical Review B 69, 220503, 2004.

[88] L. Cooper, Superconductivity in the neighborhood of metallic contacts, Physical Review Letters 6, 689 (1961).

[89] S. Adam, L. Piraux, S Michotte, D. Lucot and D. Mailly, Discontinuous hotspot growth related to the thermal healing length in superconducting NbN, Journal of Physics: Conference Series 234, 022001, 2010.

[90] F. Marsili, D. Bitauld, A. Fiore, A. Gaggero, F. Mattioli, R. Leoni, M. Benkahoul, and F. Levy, High efficiency NbN nanowire superconducting single photon detectors fabricated on MgO substrates from a low temperature process, Optics Express 16, 3191-6, 2008.

[91] F. Lefloch, C. Hoffmann, O. Demolliens, Nonlinear flux flow in TiN superconducting thin film, Physica C 319, 258, 1999.

[92] M. Yu. Kupriyanov and V. F. Lukichev, Temperature Dependence of Pair-breaking, Fiz. Nizk. Temp. 6, 445, 1980 [Sov. J. Low Temp. Phys. 6, 210, 1980].

[93] M. Hofherr, Real-time imaging systems for superconducting nanowire single-photon detector arrays, PhD thesis, Karlsruhe Insitute of Technology, 2014.

[94] J. Bardeen, L. N. Cooper, and J. R. Schrieffer, Theory of superconductivity, Physical Review 108, 1175, 1957.

[95] R. Gaudio, K. P. M. op 't Hoog, Z. Zhou, D. Sahin, and A. Fiore, Inhomogeneous critical current in nanowire superconducting single-photon detectors, Applied Physics Letters 105, 222602, 2014.

[96] D. Yu. Vodolazov, Current dependence of the red boundary of superconducting single-photon detectors in the modified hot-spot model, Physical Review B 90, 054515, 2014.

[97] Y.-J. Kim, B.J. Chun, Y. Kim, S. Hyun, and S.-W. Kim, Generation of optical frequencies out of the frequency comb of a femtosecond laser for DWDM telecommunication, Laser Physics Letters 7, No. 7, 522–527, 2010.

[98] N. Calandri, Q.-Y. Zhao, D. Zhu, A. Dane, and K. K. Berggren, Superconducting nanowire detector jitter limited by detector geometry, Applied Physics Letters 109, 152601, 2016.

[99] A. Engel, A. Semenov, H.-W. Hübers, K. Il'in, M. Siegel, Fluctuations and dark count rates in superconducting NbN single-photon detectors, Physica Status Solidi 2(5),1668–1673, 2005.

[100] L. N. Bulaevskii, M. J. Graf, and V. G. Kogan, Vortex-assisted photon counts and their magnetic field dependence in single-photon superconducting detectors, Physical Review B 85, 014505, 2012.

[101] M. Tinkham, Introduction to Superconductivity, McGraw-Hill, New York, 1996.

[102] D.Yu. Vodolazov, private communication.

[103] D.Yu. Vodolazov, Saddle point states in two-dimensional superconducting films biased near the depairing current, Physical Review B 85, 174507, 2012.

[104] K. Ilin, M. Siegel, A. Semenov, A. Engel, and H.-W. Hübers, Critical current of Nb and NbN thin-film structures: The cross-section dependence, Physica Status Solidi C 2, 1680, 2005.

[105] A. N. Zotova and D. Yu. Vodolazov, Intrinsic detection efficiency of superconducting nanowire single photon detector in the modified hot spot model, Superconductor Science and Technology 27, 125001, 2014.

[106] A. Engel and A. Schilling, Numerical analysis of detection-mechanism models of superconducting nanowire single-photon detector, Journal of Applied Physics 114, 214501, 2013.

[107] G. R. Berdiyorov, M. V. Milošević, and F. M. Peeters, Spatially dependent sensitivity of superconducting meanders as single-photon detectors, Applied Physics Letters 100, 262603, 2013.

[108] A. N. Zotova and D. Y. Vodolazov, Differences in the effects of turns and constrictions on the resistive response in current-biased superconducting wire after single photon absorption, Superconductor Science and Technology 26, 075008, 2013.

[109] P. Haas, A. Semenov, H.-W. Hübers, J. Beyer, A. Kirste, T. Schurig, K. Il'in, M. Siegel, A. Engel, and A. Smirnov, Spectral Sensitivity and Spectral Resolution of Superconducting Single-Photon Detectors, IEEE Transactions on Applied Superconductivity 17, 298, 2007.

[110] http://www.comsol.com.

[111] http://www.lumerical.com.

[112] J. J. Renema, R. J. Rengelink, I. Komen, Q. Wang, R. Gaudio, K. P. M. op 't Hoog, Z. Zhou, D. Sahin, A. Fiore, P. Kes, J. Aarts, M. P. van Exter, M. J. A. de Dood, and E. F. C. Driessen, The effect of magnetic field on the intrinsic detection efficiency of superconducting single-photon detectors, Applied Physics Letters 106(9), 2014.

List of own publications

[SCL+15] A. Semenov, **I. Charaev**, R. Lusche, K. Ilin, M. Siegel, H.-W. Huebers, N. Bralovi´c, K. Dopf, and D. Yu. Vodolazov. Asymmetry in the effect of magnetic field on photon detection and dark counts in bended nanostrips. *Phys. Rev. B 92*, 174518, 2015.

[CSL+15] **I. Charaev**, A. Semenov, R. Lusche, K. Ilin, H.-W. Huebers, M. Siegel. Enhancement of critical currents and photon-count rates by magnetic field in spiral superconducting nanowire single-photon detectors. *IEEE Trans. Applied Superconductivity 26*, 2200304, 2016.

[CSD+17] **I. Charaev**, A. Semenov, S. Doerner, G. Gomard, K. Ilin, and M. Siegel. Current dependence of the hot-spot response spectrum of superconducting single-photon detectors with different layouts. *Supercond. Sci. Technol. 30*, 025016, 2017.

[CSB+17] **I. Charaev**, T. Silbernagel, B. Bachowsky, A. Kuzmin, S. Doerner, K. Ilin, A. Semenov, D. Roditchev, D. Yu. Vodolazov, and M. Siegel. Enhancement of superconductivity in NbN nanowires by negative electron-beam lithography with positive resist. *J. Appl. Phys. 122*, 083901, 2017.

[ZEW+16] X. Zhang, A. Engel, Q. Wang, A. Schilling, A. Semenov, M. Sidorova, H.- W. Hübers, **I. Charaev**, K. Ilin, and M. Siegel. Characteristics of superconducting tungsten silicide WxSi1−x for single photon detection. *Phys. Rev. B 94*, 174509, 2016.

[DKW+17] S. Doerner, A. Kuzmin, S. Wuensch, **I. Charaev**, F. Boes, T. Zwick, and M. Siegel. Frequency-Multiplexed bias and readout of a 16-pixel Superconducting Nanowire Single-Photon Detector Array. *Appl. Phys. Lett. 111*, 032603, 2017.

[DKW++17] S. Doerner, A. Kuzmin, S. Wuensch, **I. Charaev**, and M. Siegel. Operation of Multi-Pixel Radio-Frequency Superconducting Nanowire Single-Photon Detector Arrays. *IEEE Trans. Applied Superconductivity 26*, 16594788, 2017.

[PAK+13] J. T. Peltonen, O. V. Astafiev, Yu. P. Korneeva, B. M. Voronov, A. A. Korneev, **I. A. Charaev**, A. V. Semenov, G. N. Golt'sman, L. B. Ioffe, T. M. Klapwijk, and J. S. Tsai. Coherent Flux Tunneling Through NbN Nanowires, *Phys. Rev. B 88*, 220506(R), 2013.

[FKG+13] G. Fedorov, A. Kardakova, I. Gayduchenko, **I. Charaev**, B. M. Voronov, M. Finkel, T. M. Klapwijk, S. Morozov, M. Presniakov, I. Bobrinetskiy, R. Ibragimov, and G. Goltsman. Photothermoelectric response in asymmetric carbon nanotube devices exposed to sub-THz radiation. *Appl. Phys. Lett.*, L13-07586R, 2013.

[CSB++17] **I. Charaev**, T. Silbernagel, B. Bachowsky, A. Kuzmin, S. Doerner, K. Ilin, A. Semenov, D. Roditchev, D. Yu. Vodolazov, and M. Siegel. NSN proximity effect model of ultra-narrow NbN strips. *Phys. Rev. B 96, 184517*, 2017.

[SSH+17] M. Sidorova, A. Semenov, H.- W. Hübers, **I. Charaev**, A. Kuzmin, S. Doerner, M. Siegel. Physical mechanisms of timing jitter in photon detection by current carrying superconducting nanowires. *Phys. Rev. B 96, 184504*, 2017.

[DKG+17] S. Derner, A. Kuzmin, K. Graf, S.Wuensch, **I.Charaev**, M. Siegel. The compact microwave kinetic inductance nanowire galvanometer for cryogenic detectors at 4.2 K. *J. Phys. Commun. 2,* 025016, 2018.

Conferences

1. **I. Charaev**, K. Ilin, M. Siegel. Ultra-thin TaN films on different substrates for superconducting detectors, *DPG Frühjahrstagung*, Dresden, Germany (2014)

2. **I. Charaev**, K. Ilin, M. Siegel. Single-spiral nanowire for improvement of superconducting single-photon detectors, *Tagung Kryoelektronische Bauelemente*, Berlin, Germany (2014)

3. **I. Charaev**, K. Ilin, M. Siegel. Magnetic-field enhancement of the critical current in superconducting single-spiral nanowires for broadening of the spectral range of SNSPD, *Applied Superconductivity Conference*, Charlotte, USA (2014)

4. **I. Charaev**, K. Ilin, M. Siegel. Magnetic-field enhanced superconductivity in mesoscopic structures, *DPG Frühjahrstagung*, Berlin, Germany (2015)

5. **I.Charaev**, R. Lusche, A. Semenov, K. Il'in, M. Siegel. Critical currents and detection efficiency of the spiral SNSPDs in magnetic field, *12th European Conference on Applied Superconductivity*, Lyon, France (2015)

6. **I.Charaev**, R. Lusche, A. Semenov, K. Il'in, M. Siegel. Single-spiral superconducting nanowire single-photon detector in magnetic field, *Tagung Kryoelektronische Bauelemente*, Warberg, Germany (2015)

7. **I. Charaev**, K. Ilin, M. Siegel. Single-spiral superconducting nanowire single-photon detector, *Karlsruhe Days of Optics and Photonics*, Karlsruhe, Germany (2015)

8. **I.Charaev**, R. Lusche, A. Semenov, K. Il'in, M. Siegel. Study of single-spiral superconducting nanowire single-photon detectors in magnetic fields, *DPG Frühjahrstagung*, Regenburg, Germany (2016)

9. **I. Charaev**, A. Semenov, K. Ilin, M. Siegel. Spectra of detection efficiency of single-spiral superconducting nanowire single-photon detectors in magnetic field, *Applied Superconductivity Conference*, Colorado, USA (2016)

10. **I. Charaev**, A. Semenov, S. Doerner, G. Gomard, K. Ilin, M. Siegel. Current dependence of the hot-spot response spectrum of superconducting single-photon detectors with different layouts, *European Conference on Applied Superconductivity*, Geneva, Switzerland (2017)

11. **I. Charaev**, A. Semenov, S. Doerner, G. Gomard, K. Ilin, M. Siegel. Current dependence of the hot-spot response spectrum of superconducting single-photon detectors with different layouts, *Tagung Kryoelektronische Bauelemente*, Bad Aibling, Germany (2017)

List of Figures

List of Tables

Nomenclature

A_{max}	Maximal pulse amplitude
ABS	Absorption efficiency
a	Constant parameter, gap between two lines
B	Magnetic field
B_{c2}	Second critical magnetic field
B_{edge}	Magnetic field generated by transport current at the edge of the strip
β_0	Ratio of the energy gap at zero temperature to $k_B T_C$
CIPD	Charge integration photon detector
CVC	Current-voltage characteristic
C	Constant
c	Speed of light
D	Quasiparticle diffusion coefficient
DCR	Dark count rate
DE	Detection efficiency
d	Thickness
d_N	Thickness of normal conducting surface layer
dB/dT	Derivative of the second critical magnetic field
Δ	Superconducting energy gap
ΔT	Temperature span of superconducting transition
ΔW	Width of the damaged bands
δF	Local height of the barrier

δI	Local critical current
δ	Local decrease of the order parameter
E	Photon energy
e	Elementary charge
ξ	Coherent length
ς	Conversion efficiency of the photon from initial excitation in the film to quasiparticles at the superconducting gap
FWHM	Full width at half maximum
ψ^2	Order parameter
Φ_0	Magnetic flux quantum
GL	Ginzburg-Landau
σ_{system}	System jitter
σ_{SNSPD}	Intrinsic jitter
σ_{noise}	Jitter of readout circuit
σ_{N}	Root mean square noise amplitude
HTS	High-temperature superconductor
h	Planck constant
h	Reduced Planck's constant
IDE	Intrinsic detection efficiency
I_{b}	Bias current
I_{C}	Critical current
$I_{\mathrm{C}}^{\mathrm{d}}$	Depairing current limit
I_{det}	Detection current

I_{dis}	Discharge current
$I_{c,v}$	Critical current that reduces the energy barrier for vortex penetration to zero
I_0	Bias current (QDT)
I_r	Retrapping current
j_C	Critical current density
j_r	Density of retrapping current
KL	Correction factor of Kupriyanov and Lukichev
κ	Curvature of spiral
k_B	Boltzmann's constant
L	Length
ℓ	Electron mean free path, length of the nanowire
λ	Wavelength
λ_C	Cut-off wavelength
$\lambda(0)$	Penetration depth
Λ	Pearl length
μ_0	Vacuum permeability
N	Length of segment
NSN	Normal conductor-Superconductor-Normal conductor
N_0	Density of electron states
n_{qp}	Number of quasiparticles
ν	Frequency of radiation, vortex energy scale in the initial state, Fermi velocity, group velocity
OCE	Optical coupling efficiency

QDT	Quantum detector tomography
PCB	Printed circuit board
PCR	Photon count rate
P_{Ar}	Partial pressure of argon gas
P_{N2}	Partial pressure of nitrogen gas
p	Exponent
RRR	Residual resistivity ratio
R	Reduction factor of current crowding effect
$R_{(X)}$	Resistance (at temperature x)
R_{sq}	Square resistance
r	Reflectance, radial distance, nominal rounding of inner corner
$\rho(x)$	Resistivity (at temperature x)
S	Surface
SDE	System detection efficiency
SE	Secondary electrons
SEM	Scanning Electron Microscope
SNSPD	Superconducting Nanowire Single-Photon Detector
SPAD	Single-photon avalanche photodiode
τ	Scattering/ relaxation time constant
τ_{rise}	Pulse rise time
T	Temperature
T_{C}	Critical temperature
$T_{\mathrm{C}}^{\mathrm{CP}}$	Critical temperature of contact pads

t	Transmittance, reduced temperature
t_o	Temperature of the interleaving
U	Voltage
θ	Polar angle
V	Volume
VAP	Vortex-antivortex pair
W	Width
W_{SEM}	Nominal width
W_{SC}	Effective width of the superconducting core
γ	Material and device dependent proportionality factor

Karlsruher Schriftenreihe zur Supraleitung

Karlsruher Institut für Technologie (KIT) | ISSN 1869-1765

Herausgeber: Prof. Dr.-Ing. M. Noe, Prof. Dr. rer. nat. M. Siegel

Band 001 **Christian Schacherer**
Theoretische und experimentelle Untersuchungen zur Entwicklung supraleitender resistiver Strombegrenzer. 2009
ISBN 978-3-86644-412-6

Band 002 **Alexander Winkler**
Transient behaviour of ITER poloidal field coils. 2011
ISBN 978-3-86644-595-6

Band 003 **André Berger**
Entwicklung supraleitender, strombegrenzender Transformatoren. 2011
ISBN 978-3-86644-637-3

Band 004 **Christoph Kaiser**
High quality Nb/Al-AlOx/Nb Josephson junctions. Technological development and macroscopic quantum experiments. 2011
ISBN 978-3-86644-651-9

Band 005 **Gerd Hammer**
Untersuchung der Eigenschaften von planaren Mikrowellen-resonatoren für Kinetic-Inductance Detektoren bei 4,2 K. 2011
ISBN 978-3-86644-715-8

Band 006 **Olaf Mäder**
Simulationen und Experimente zum Stabilitätsverhalten von HTSL-Bandleitern. 2012
ISBN 978-3-86644-868-1

Die Bände sind unter www.ksp.kit.edu als PDF frei verfügbar oder als Druckausgabe bestellbar.

Band 007 **Christian Barth**
High Temperature Superconductor Cable Concepts for Fusion Magnets. 2013
ISBN 978-3-7315-0065-0

Band 008 **Axel Stockhausen**
Optimization of Hot-Electron Bolometers for THz Radiation. 2013
ISBN 978-3-7315-0066-7

Band 009 **Petra Thoma**
Ultra-fast $YBa_2Cu_3O_{7-x}$ direct detectors for the THz frequency range. 2013
ISBN 978-3-7315-0070-4

Band 010 **Dagmar Henrich**
Influence of Material and Geometry on the Performance of Superconducting Nanowire Single-Photon Detectors. 2013
ISBN 978-3-7315-0092-6

Band 011 **Alexander Scheuring**
Ultrabreitbandige Strahlungseinkopplung in THz-Detektoren. 2013
ISBN 978-3-7315-0102-2

Band 012 **Markus Rösch**
Development of lumped element kinetic inductance detectors for mm-wave astronomy at the IRAM 30 m telescope. 2013
ISBN 978-3-7315-0110-7

Band 013 **Johannes Maximilian Meckbach**
Superconducting Multilayer Technology for Josephson Devices. 2013
ISBN 978-3-7315-0122-0

Band 014 **Enrico Rizzo**
Simulations for the optimization of High Temperature Superconductor current leads for nuclear fusion applications. 2014
ISBN 978-3-7315-0132-9

Die Bände sind unter www.ksp.kit.edu als PDF frei verfügbar oder als Druckausgabe bestellbar.

Band 015 **Philipp Krüger**
Optimisation of hysteretic losses in high-temperature superconducting wires. 2014
ISBN 978-3-7315-0185-5

Band 016 **Matthias Hofherr**
Real-time imaging systems for superconducting nanowire single-photon detector arrays. 2014
ISBN 978-3-7315-0229-6

Band 017 **Oliver Näckel**
Development of an Air Coil Superconducting Fault Current Limiter. 2016
ISBN 978-3-7315-0526-6

Band 018 **Christoph M. Bayer**
Characterization of High Temperature Superconductor Cables for Magnet Toroidal Field Coils of the DEMO Fusion Power Plant. 2017
ISBN 978-3-7315-0605-8

Band 019 **Shengnan Zou**
Magnetization of High Temperature Superconducting Trapped-Field Magnets. 2017
ISBN 978-3-7315-0715-4

Band 020 **Ilya Charaev**
Improving the Spectral Bandwidth of Superconducting Nanowire Single-Photon Detectors (SNSPDs). 2018
ISBN 978-3-7315-0745-1